Spatial Control of Vibration

Theory and Experiments

SERIES ON STABILITY, VIBRATION AND CONTROL OF SYSTEMS

About the Series

Rapid developments in system dynamics and control, areas related to many other topics in applied mathematics, call for comprehensive presentations of current topics. This series contains textbooks, monographs, treatises, conference proceedings and a collection of thematically organized research or pedagogical articles addressing dynamical systems and control.

The material is ideal for a general scientific and engineering readership, and is also mathematically precise enough to be a useful reference for research specialists in mechanics and control, nonlinear dynamics, and in applied mathematics and physics.

Selected Volumes in Series B

Proceedings of the First International Congress on Dynamics and Control of Systems, Chateau Laurier, Ottawa, Canada, 5–7 August 1999
Editors: A. Guran, S. Biswas, L. Cacetta, C. Robach, K. Teo, and T. Vincent

Selected Volumes in Series A

Vol. 1 Stability Theory of Elastic Rods
Author: T. Atanackovic

Vol. 2 Stability of Gyroscopic Systems
Authors: A. Guran, A. Bajaj, Y. Ishida, G. D'Eleuterio, N. Perkins, and C. Pierre

Vol. 3 Vibration Analysis of Plates by the Superposition Method
Author: Daniel J. Gorman

Vol. 4 Asymptotic Methods in Buckling Theory of Elastic Shells
Authors: P. E. Tovstik and A. L. Smirinov

Vol. 5 Generalized Point Models in Structural Mechanics
Author: I. V. Andronov

Vol. 6 Mathematical Problems of Control Theory: An Introduction
Author: G. A. Leonov

Vol. 7 Analytical and Numerical Methods for Wave Propagation in Fluid Media
Author: K. Murawski

Vol. 8 Wave Processes in Solids with Microstructure
Author: V. I. Erofeyev

Vol. 9 Amplification of Nonlinear Strain Waves in Solids
Author: A. V. Porubov

SERIES ON STABILITY, VIBRATION AND CONTROL OF SYSTEMS

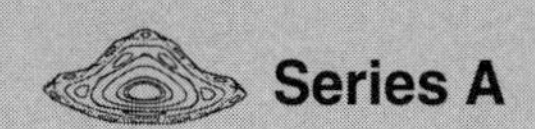
Series A Volume 10

Spatial Control of Vibration

Theory and Experiments

S. O. Reza Moheimani
The University of Newcastle, Australia

Dunant Halim
Adelaide University, Australia

Andrew J. Fleming
The University of Newcastle, Australia

NEW JERSEY • LONDON • SINGAPORE • SHANGHAI • HONG KONG • TAIPEI • BANGALORE

Published by

World Scientific Publishing Co. Pte. Ltd.
5 Toh Tuck Link, Singapore 596224
USA office: Suite 202, 1060 Main Street, River Edge, NJ 07661
UK office: 57 Shelton Street, Covent Garden, London WC2H 9HE

British Library Cataloguing-in-Publication Data
A catalogue record for this book is available from the British Library.

SPATIAL CONTROL OF VIBRATION: THEORY AND EXPERIMENTS

ISBN 981-238-337-9

Printed in Singapore by World Scientific Printers (S) Pte Ltd

To Our Parents

Preface

A growing trend in the manufacturing of engineering systems has been to reduce the weight of the structure. For certain engineering systems, such as airplanes, spacecraft, etc. this would result in a substantial reduction of operating costs. However, as a direct consequence of this the structure will exhibit more flexible dynamics. The addition of such flexible modes to the rigid dynamics of the structure can limit the performance of the system since the structure will be more susceptible to vibrations. The traditional method of treatment is to add damping to the system using passive techniques. Passive methods are known to be effective at higher frequencies. However, they are hardly useful in dealing with low frequency vibrations.

Active vibration control has been under investigation as an alternative tool for bridging this low frequency gap. Two general frameworks for vibration control have been proposed. These are the adaptive feedforward control, and the feedback control methods. In both techniques a sensor (or an array of sensors) is used to measure vibration of the structure, and a control signal is applied to an actuator (or an array of actuators) to minimize structural vibrations. A problem that could arise is the so-called "spatial spill-over effect". Although the vibration level at, and in the vicinity of the sensors is reduced due to the control action, vibration of certain areas on the structure could be adversely affected once the controller is activated. This is due to the spatially distributed nature of flexible structures which implies that vibrations of distinct points are dynamically related.

This monograph introduces the concept of "spatial control". Although flexible structures are spatially distributed systems, the sensors and actuators that are used for active vibration control purposes are discrete*.

*In fact they may be spatially distributed, e.g. piezoelectric transducers. However, the control and measurement signals are viewed as discrete signals.

Therefore, a spatially distributed system is often controlled using a number of discrete actuators and sensors. The spatial information embedded in the dynamical model of a structure can be used to set up spatial cost functions. A controller can then be designed with a view to stabilizing the system while minimizing an appropriate spatial cost function. This would result in controllers capable of reducing structural vibrations in a spatial sense.

The material presented in this monograph reflects several years of research dating back to 1996 when the first author was a postdoctoral research fellow at the Australian Defence Force Academy. In 1997 he joined the Department of Electrical and Computer Engineering[†] at the University of Newcastle, where he established the *Laboratory for Dynamics and Control of Smart Structures*. The second and third co-authors subsequently joined the group as graduate students.

In the production of this book, the authors wish to acknowledge the support they have received from the Australian Research Council (ARC) and the Defence Science and Technology Organization (DSTO). Also, the authors wish to acknowledge the contributions made by their colleagues, Sam Behrens, Will Heath, Ian Petersen, Hemanshu Pota and Tom Ryall, in the research that underlies parts of the material presented in this book. Furthermore, the first author is grateful for the enormous support he has received from his wife, Niloofar.

S. O. R. Moheimani
D. Halim
A. J. Fleming

Newcastle, December 2002

[†]Now a part of School of Electrical Engineering and Computer Science.

Contents

Chapter 1

Introduction

1.1 Vibration

Vibration is a natural phenomenon that occurs in many engineering systems. In many circumstances, vibration greatly affects the nature of engineering design as it often dictates limiting factors in the performance of the system. The conventional method of treatment is to redesign the system or to use passive damping. The former could be a costly exercise, while the latter is only effective at higher frequencies. Active control techniques have emerged as viable technologies to fill this low-frequency gap.

This book is concerned with the study of feedback controllers for vibration suppression in flexible structures, with a view to minimizing vibration over the entire body of the structure. Following, is an overview of some of the topics covered in this monograph.

1.2 Spatially distributed systems

A large number of systems can be classified as "spatially distributed systems". This monograph is mainly concerned with spatially distributed systems that arise in vibration control applications, i.e. flexible structures. Chapter 2 introduces a number of such systems, clarifying the process of physical modeling.

The dynamics of a system, such as a beam, plate, or string, are governed by a specific class of partial differential equations. Such a partial differential equation can be discretized, using the modal analysis procedure, to obtain a lumped, but spatially distributed model for the system under consideration. An eigenvalue problem must be solved to obtain "mode shapes" that describe spatial characteristics of the system [67, 68].

This procedure can be quite successful, as long as the boundary conditions associated with the describing partial differential equation are such that the resulting eigenvalue problem can be solved analytically. Chapter 2 contains a number of examples, such as beams and plates with pinned boundary conditions, for which analytic solutions to the associated eigenvalue problems can be found.

For the majority of realistic flexible structures, the above procedure becomes difficult or impossible, thus failing to produce a model for the system. The problem gets significantly complicated for irregular structural geometries and/or unknown or non-trivial boundary conditions. To obtain a spatially distributed model, one may have to resort to Finite Element Methods [93, 10, 24]. Such methods can become time-consuming or inaccurate as the procedure requires intimate knowledge of the system's physical properties. Furthermore, the finite element technique results in models of very high orders, unsuitable for control design purposes.

A number of *ad hoc* methods are being employed by practicing engineers to generate spatially distributed models from experimental data obtained from flexible systems. An overview of these techniques can be found in [45]. These are crude, yet perhaps effective methods that often result in models that provide the engineer with some insight into the spatial, as well as spectral dynamics of the underlying system. For control design purposes, such models tend to be rather inaccurate.

A viable option is to use a system identification method to procure a spatially distributed model from experimental data. Such a technique is presented in Chapter 7. Based on a number of measurements over the surface of the structure, a multi-variable model is identified. This part of the modeling is carried out using a frequency domain subspace-based system identification technique [65, 60, 61, 109, 66, 64] . Subspace-based system identification techniques have proved to be effective methods for identifying multi-variable, lightly damped systems of high orders. A spline reconstruction method [42, 53, 107] is then used to construct a spatially distributed, though finite-dimensional in state, model of the system from experimental data.

1.3 Model correction

The spatially distributed systems considered in this monograph are of a particular form. They have a finite-dimensional state vector, however, their output is spatially distributed over a specific set. This property results in controller design problems that are tractable since finite-dimensional controllers can be designed to achieve required performance as measured by a spatial performance index.

An accurate model of the underlying system representing the spatial, as well as spectral dynamics of the system, is required for spatial controller design. The desired performance specification dictates the requisite model accuracy.

If a system is modeled via modal analysis [67], the resulting system will contain an infinite number of second order, highly resonant terms. To obtain a finite-dimensional model, high-frequency terms that lie out of the bandwidth of interest need to be truncated. Due to the influence of high-frequency dynamics on in-bandwidth zeros, model truncation may generate significant uncertainty. If the controller is designed using the truncated model, this may have an adverse effect on the closed-loop performance of the system.

This problem can be alleviated by adding a feed-through term to the truncated model to account for the loss of low-frequency information resulting from truncation of out-of-bandwidth modes. For a spatially distributed system, the resulting feed-through term will be a function of the spatial variables. A number of methods for obtaining the correcting feed-through term have been suggested in the literature [11, 77, 72, 70, 71, 75]. Some of these techniques are described in Chapter 4.

To study the class of spatially distributed systems that are the subject of this book, it is necessary to develop mathematical machinery capable of addressing different issues that arise in analysis and control synthesis problems associated with such systems. Chapter 3 introduces a number of measures such as spatial $\mathcal{H}_2$ and spatial $\mathcal{H}_\infty$ norms for the class of spatially distributed systems considered in this monograph.

1.4 Spatial control

Vibration of every point on the surface of a flexible structure is dynamically related to that of every other point via a well-defined transfer function. Such a spatially distributed system is often controlled using point-wise sensors and actuators incapable of producing spatial measurements and actuation. This is largely due to technological constraints.

One of the main arguments in this book is that in controlling spatially distributed systems with point-wise actuators and sensors, the controller must guarantee a certain level of performance with respect to a spatial cost function. In particular, Chapter 5 demonstrates that if a controller is designed with a view to minimizing vibration at a limited number of points, the resulting controller may perform poorly, in the spatial sense, when implemented on the structure. Recently, similar observations have been made by other researchers in the context of active noise control in ducts [104, 105].

The solution proposed in this book is to employ spatial cost functions by re-defining classical $\mathcal{H}_2$ and $\mathcal{H}_\infty$ control problem formulations as spatial $\mathcal{H}_2$ and $\mathcal{H}_\infty$ control problems (see also [34, 35, 37, 38]). In Chapter 5, these control design techniques are applied to a simply-supported beam and, through experiments, their effectiveness is demonstrated.

1.5 Piezoelectric actuators and sensors

One of the features of this monograph is that almost all of the proposed methodologies are tested through experiments on flexible structures. The experimental test-beds comprise of flexible beams and plates, with well-defined boundary conditions. Piezoelectric transducers are used as feedback sensors and actuators.

Piezoelectric transducers are typically used in applications that require the transformation of electrical energy to mechanical energy or *vice versa*. Other design considerations include: size limitations, precision, and speed of operation. During the past decade, there has been a great deal of interest in the use of piezoelectric sensors and actuators for structural control applications. In particular, piezoelectric elements have been used successfully in the closed loop control of a variety of active structures including beams, plates, shafts, and trusses; see [57, 106, 100] and references therein. Also, they have been used in the suppression of flutter in panels [21], lifting surfaces [54], hard disk drive heads [94] and airfoils [55]. Piezoelectric ceramics provide cheap, reliable and non-intrusive means of actuation and sensing in flexible structures. A structure that is constructed by integration of piezoelectric actuators and/or sensors with a flexible body is often referred to as a *smart structure*, an *adaptive structure*, or an *active structure*. Smart structures are structures that are self-sensing and self-compensating.

Two classes of piezoelectric materials are currently being used. These are poly-vinylidene fluoride (PVDF), a semi-crystalline polymer film, and lead zirconate titanate (PZT), a piezoelectric ceramic material. Both PVDF and PZT materials have the property that they strain when exposed to a voltage, and conversely produce a voltage when mechanically strained. Experiments reported in this book were performed using PZT-based transducers. PZT materials have the advantageous property of having a higher coupling factor between the applied electric field and the resulting mechanical strain, and therefore, require a lower level of voltage than is needed in PVDF films to produce acceptable actuation.

1.6 Actuator and sensor placement

Due to the spatially distributed nature of flexible structures, the actuators and sensors can be placed in many locations. It is, therefore, natural to ask questions such as where should one place these actuators and sensors to obtain higher levels of performance. A number of researchers have tackled this problem; see [27, 14, 43, 20, 17].

It appears that there are two main approaches to the problem of actuator and sensor placement in flexible structures. One approach is based on integrating the problem of actuator and/or sensor placement with a specific control design methodology, such as LQG, and treating the locations of actuators and sensors as extra design parameters. Another approach treats the question of placement of actuators and sensors independently from the control design problem with the view that once actuators and sensors are positioned at suitable locations on the structure, a wide range of control design techniques can be employed for minimizing structural vibrations. The latter approach is adopted in this monograph.

The problem of actuator and sensor placement is discussed in Chapter 6, where a solution to the problem is given based on the notions of "spatial controllability and observability" and "modal controllability and observability" (see also [74, 33]). The proposed methodology is validated experimentally on a piezoelectric laminate plate with simply-supported boundary conditions.

Chapter 2

Modeling

2.1 Introduction

This book is concerned with spatially distributed systems. Therefore, it is natural to start it with a chapter on modeling physical systems that are classified as spatially distributed.

The main purpose of the book is to develop control design methodologies for vibration control of flexible structures with a view to minimizing vibration everywhere. Therefore, in this chapter, we restrict our attention to developing dynamic models for flexible structures. In doing so, we pay particular attention to obtaining models that describe spatial characteristics of the structures. It needs to be pointed out that there are many other engineering systems that can be classified as spatially distributed, e.g. acoustic ducts and enclosures [41]. Although the methodologies developed in this book may be applicable to other spatially distributed systems, those systems are not discussed any further, in the interest of staying focused. In modeling the structures considered in this chapter, an important assumption is that the deformation of the systems is kept sufficiently small. In other words, the angle or slope that a structure makes with respect to its undisturbed position is small enough so that a small angle approximation can be employed. This approximation simplifies the overall derivation and allows the equations of motion to be linearized.

2.2 Modal approach

In this section, we review the mathematical basis upon which a class of spatio-temporal systems can be modeled. The presentation is made as general as possible within the context of systems that are of importance to us, while the forthcoming sections will concentrate on specific structures that fit withing this framework. We consider a partial differential equation described by:

$$\mathcal{L}\left\{y(t,r)\right\} + \mathcal{M}\left\{\frac{\partial^2 y(t,r)}{\partial t^2}\right\} = f(t,r). \tag{2.1}$$

Here, r is defined over a domain $\mathcal{R}$, $\mathcal{L}$ is a linear homogeneous differential operator of order $2p$, $\mathcal{M}$ is a linear homogeneous differential operator of order $2q$, $q \leq p$ and $f(t,r)$ is the system input, which could be spatially distributed over $\mathcal{R}$. Corresponding to this partial differential equation are the following boundary conditions:

$$\mathcal{B}_\ell\left\{y(t,r)\right\} = 0, \qquad \ell = 1, 2, \ldots, p. \tag{2.2}$$

These boundary conditions are to be satisfied at every point of the boundary $\mathcal{S}$ of the domain $\mathcal{R}$. Here, $\mathcal{B}_\ell$, $\ell = 1, 2, \ldots, p$ are linear differential operators of orders ranging from zero to $2p-1$. We notice that (2.1) and (2.2) describe *spatial* and *temporal* behavior of y. Next we explain how a model can be derived that captures the spatial and temporal characteristics of (2.1) and (2.2). The modal analysis is concerned with seeking a solution for (2.1) in the form

$$y(t,r) = \sum_{i=1}^{\infty} \phi_i(r) q_i(t). \tag{2.3}$$

Here $\phi_i(\cdot)$ are the eigenfunctions that are obtained by solving the eigenvalue problem associated with (2.1). That is,

$$\mathcal{L}\{\phi_i(r)\} = \lambda_i \mathcal{M}\{\phi_i(r)\}$$

and its associated boundary conditions,

$$\mathcal{B}_\ell\{\phi_i(r)\} = 0, \quad \ell = 1, 2, \ldots, p, \quad i = 1, 2, \ldots \ .$$

The solution of the eigenvalue problem consists of an infinite set of eigenvalues λ_i, $i = 1, 2, \ldots$ and associated eigenfunctions $\phi_i(r)$. Assuming that the operator $\mathcal{L}$ is self-adjoint and positive definite, all the eigenvalues will be positive and can be ordered so that

$$\lambda_1 \leq \lambda_2 \leq \ldots \ .$$

Moreover, the eigenvalues are related to the natural frequencies of the system via

$$\lambda_i = \omega_i^2, \quad i = 1, 2, \ldots \quad .$$

In the modal analysis literature, ϕ_is are often referred to as mode shapes. Since $\mathcal{L}$ is self-adjoint, the mode shapes possess the orthogonality property and are normalized via the following orthogonality conditions:

$$\int_{\mathcal{R}} \phi_i(r)\mathcal{L}\{\phi_j(r)\}dr = \delta_{ij}\omega_i^2 \tag{2.4}$$

$$\int_{\mathcal{R}} \phi_i(r)\mathcal{M}\{\phi_j(r)\}dr = \delta_{ij}, \tag{2.5}$$

where δ_{ij} is the Kronecker delta function, i.e., $\delta_{ij} = 1$ for $i = j$, and zero otherwise. To this end we point out that the expansion theorem [67] states that the series (2.3) will converge to the solution of (2.1) at every time and at every point in the domain $\mathcal{R}$. In mathematical literature, this is referred to as the Sturm-Liouville theorem [110]. Substituting (2.3) in (2.1), we obtain

$$\mathcal{L}\left\{\sum_{i=1}^{\infty} \phi_i(r)q_i(t)\right\} + \mathcal{M}\left\{\frac{\partial^2}{\partial t^2}\sum_{i=1}^{\infty} \phi_i(r)q_i(t)\right\} = f(t,r). \tag{2.6}$$

Multiplying both sides of (2.6) by $\phi_j(r)$, integrating over the domain $\mathcal{R}$ and taking advantage of the orthogonality conditions (2.4) and (2.5), we obtain an infinite number of decoupled second order, ordinary differential equations:

$$\ddot{q}_i(t) + \omega_i^2 q_i(t) = Q_i(t), \qquad i = 1, 2, \ldots, \tag{2.7}$$

where

$$Q_i(t) = \int_{\mathcal{R}} \phi_i(r) f(t,r)dr.$$

In many cases, $Q_i(t)$ can be written as

$$Q_i(t) = F_i u(t),$$

where $u(t)$ is the input of the system. That is, $f(t,r)$ can be decomposed into its spatial and temporal components. Taking the Laplace transform of (2.7), we obtain the input-output description of the system dynamics in terms of a transfer function:

$$G(s,r) = \sum_{i=1}^{\infty} \frac{\phi_i(r)F_i}{s^2 + \omega_i^2}. \tag{2.8}$$

In the rest of this chapter, we will discuss a number of systems whose dynamics are governed by partial differential equations of the form (2.1).

2.3 Transverse vibration of strings

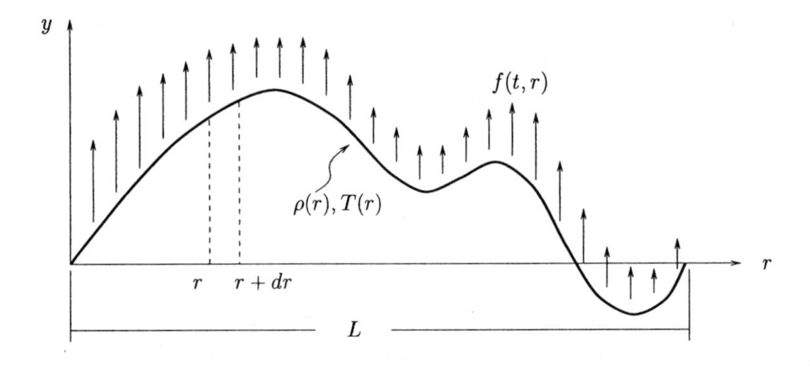

Fig. 2.1 A string in transverse vibration

A string (or a cable) is a spatially distributed system, whose lateral dimensions are significantly smaller than its longitudinal dimensions. A string is only able to support tension forces at its cross sections.

Consider a typical string of length L depicted in Figure 2.1. It is assumed that the motion is in transverse direction only. The transverse vibration of the string is defined by $y(t,r)$. The tension of the string, density, cross-sectional area and distributed force at point r are denoted by $T(r), \rho(r), A(r)$ and $f(t,r)$ respectively. Consider the forces acting on an element of the string as shown in Figure 2.2.

Applying Newton's second law for the force component along the y axis,

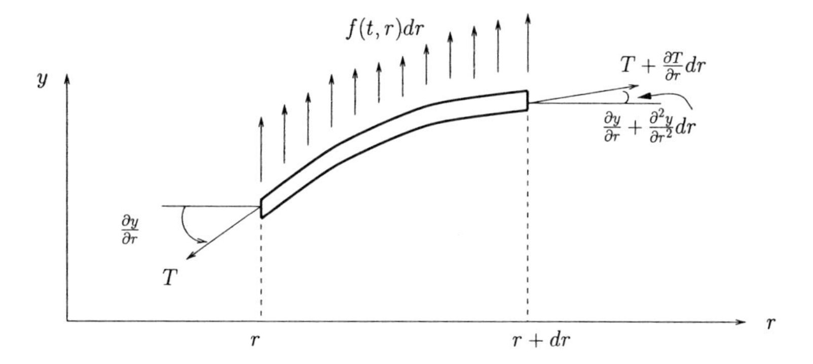

Fig. 2.2 A small element of the string

we obtain:

$$\left(T(r) + \frac{\partial T(r)}{\partial r}dr\right)\left(\frac{\partial y(t,r)}{\partial r} + \frac{\partial^2 y(t,r)}{\partial r^2}dr\right)$$
$$-T(r)\frac{\partial y(t,r)}{\partial r} + f(t,r)dr = \rho A(r)dr\frac{\partial^2 y(t,r)}{\partial t^2}. \tag{2.9}$$

Note that $\rho A(r)$ is the brief form of $\rho(r)A(r)$. We then obtain the partial differential equation (PDE) for transverse vibration of strings by simplifying the equation (2.9) and ignoring the second-order term in dr [67, 16]:

$$\frac{\partial}{\partial r}\left(T(r)\frac{\partial y(t,r)}{\partial r}\right) + f(t,r) = \rho A(r)\frac{\partial^2 y(t,r)}{\partial t^2}, \qquad 0 \le r \le L. \tag{2.10}$$

2.3.1 *A uniform string with a constant tension*

Consider the particular case of a uniform string with a constant tension, $\rho(r) = \rho$ and $T(r) = T$. Then, the PDE of the string (2.10) becomes:

$$T\frac{\partial^2 y(t,r)}{\partial r^2} + f(t,r) = \rho A\frac{\partial^2 y(t,r)}{\partial t^2}, \qquad 0 \le r \le L. \tag{2.11}$$

For the case of free vibration, i.e. $f(t,r) = 0$, the following can be obtained:

$$\frac{\partial^2 y(t,r)}{\partial r^2} = \frac{1}{c^2}\frac{\partial^2 y(t,r)}{\partial t^2}, \tag{2.12}$$

where $c = \sqrt{T/(\rho A)}$.

It can be observed that (2.12) is the wave equation [67, 16, 85, 44] whose solution consists of two traveling waves moving with the wave speed of c in forward and backward directions respectively. The solution can be determined from the initial conditions and boundary conditions of the string.

Notice that the partial differential equation (2.11) is of the same form as (2.1) with

$$\mathcal{L} = -\frac{\partial^2}{\partial r^2}$$

and

$$\mathcal{M} = \rho A.$$

Therefore, a modal solution of the form

$$y(t,r) = \sum_{i=1}^{\infty} \phi_i(r) q_i(t)$$

can be sought. The eigenfunction, $\phi_i(r)$, satisfies the associated eigenvalue problem and boundary conditions.

Some common boundary conditions for strings are as follows [16]:

- **Fixed end at $r = r_o$:** The deflection is zero.

$$y(t,r_o) = 0. \tag{2.13}$$

- **Free end at $r = r_o$:** The vertical component of the tension is zero since there is no restriction on the end's vertical translation.

$$T\frac{\partial y(t,r)}{\partial r}\bigg|_{r=r_o} = 0. \tag{2.14}$$

- **Flexible end at $r = r_o$ with linear stiffness of k:** The vertical component of the tension is equal to the equivalent spring force.

$$T\frac{\partial y(t,r)}{\partial r}\bigg|_{r=r_o} = ky(t,r_o). \tag{2.15}$$

- **Flexible and inertial end at $r = r_o$ with linear stiffness of k and mass m:** The overall force due to the string tension and equivalent spring force is equal to the inertia force due to the mass m.

$$T\frac{\partial y(t,r)}{\partial r}\bigg|_{r=r_o} - ky(t,r_o) = m\frac{\partial^2 y(t,r)}{\partial t^2}\bigg|_{r=r_o}. \tag{2.16}$$

Next, we consider an example of the modal solution for string vibration. Let us consider a fixed uniform string with the length L, i.e. the string is fixed at $r = 0$ and $r = L$. The equation for free vibration of the string is given in (2.12). The solution is assumed to be in separable form of

$$y(t, r) = \phi(r)q(t). \tag{2.17}$$

Substituting (2.17) into (2.12), we obtain:

$$\phi''(r)q(t) = \frac{1}{c^2}\phi(r)\ddot{q}(t). \tag{2.18}$$

Here $c = \sqrt{T/(\rho A)}$ and $\phi''(r)$ represents the second derivative of $\phi(r)$ with respect to r, and $\ddot{q}(t)$ is the second time derivative of $q(t)$. After re-arranging (2.18), it can be shown that:

$$\frac{\phi''(r)}{\phi(r)} = \frac{1}{c^2}\frac{\ddot{q}(t)}{q(t)} = -\lambda^2 \tag{2.19}$$

since the first term is independent of t and the second term is independent of r, both terms must be equal to a constant, $-\lambda^2$.

For a harmonic motion with frequency ω, $q(t)$ can be described in exponential form, i.e. $q(t)$ is proportional to $e^{j\omega t}$. Hence, the following result is obvious:

$$\frac{\ddot{q}(t)}{q(t)} = -\omega^2. \tag{2.20}$$

Considering (2.19) and (2.20), we obtain:

$$\lambda^2 = \frac{\omega^2 \rho A}{T}. \tag{2.21}$$

The spatial function $\phi(r)$ is assumed to be of the form [67, 16]:

$$\phi(r) = \alpha \sin \lambda r + \beta \cos \lambda r. \tag{2.22}$$

Now, the fixed boundary conditions in (2.13) require the deflections at two ends to be zero:

$$\begin{aligned} y(t, 0) &= 0, \\ y(t, L) &= 0. \end{aligned} \tag{2.23}$$

The first boundary condition requires $\beta = 0$, while the second condition yields the following:

$$\sin \lambda L = 0 \tag{2.24}$$

which has an infinite number of solutions:

$$\lambda_i = \frac{i\pi}{L}, \qquad i = 1, 2, \dots \ . \tag{2.25}$$

From (2.21), the expression for the natural frequency of mode i is

$$\omega_i = i\pi\sqrt{\frac{T}{\rho A L^2}}, \qquad i = 1, 2, \dots \ . \tag{2.26}$$

Therefore, the solution for $\phi(r)$ consists of an infinite number of eigenfunctions or mode shape functions:

$$\phi_i(r) = \alpha_i \sin\frac{i\pi r}{L}. \tag{2.27}$$

Thus, after normalizing the mode shape function, i.e. $\alpha_i = 1$, the solution $y(t,r)$ is found to be

$$\begin{aligned} y(t,r) &= \sum_{i=1}^{\infty} \sin\frac{i\pi r}{L} q_i(t) \\ \omega_i &= i\pi\sqrt{\frac{T}{\rho A L^2}}, \qquad i = 1, 2, \dots \end{aligned} \tag{2.28}$$

The first three mode shapes of the vibrating fixed uniform string are shown in Figure 2.3.

2.4 Axial vibration of rods

Consider a rod of length L depicted in Figure 2.4. The distributed axial force, density, cross-sectional area, and Young's modulus of elasticity at point r are defined as $f(t,r), \rho(r), A(r)$ and $E(r)$ respectively. The longitudinal displacement at point r is denoted by $u(t,r)$.

The main assumptions used are:

(i) The material follows Hooke's law, i.e. $\sigma(r) = \epsilon(r)E(r)$, where σ and ϵ are the longitudinal stress and strain at point r respectively [29, 90, 103].

(ii) The rod's lateral dimensions are sufficiently smaller than its longitudinal dimension so that the radial motion can be neglected.

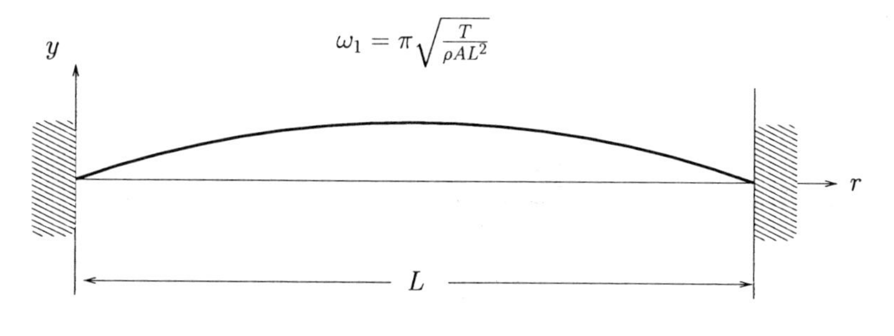

(a) Mode 1

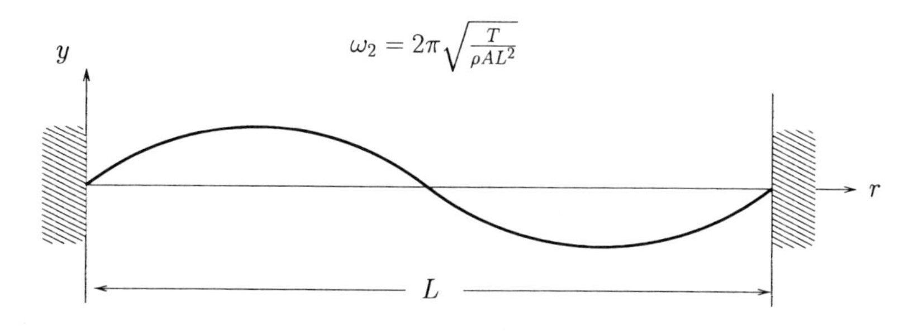

(b) Mode 2

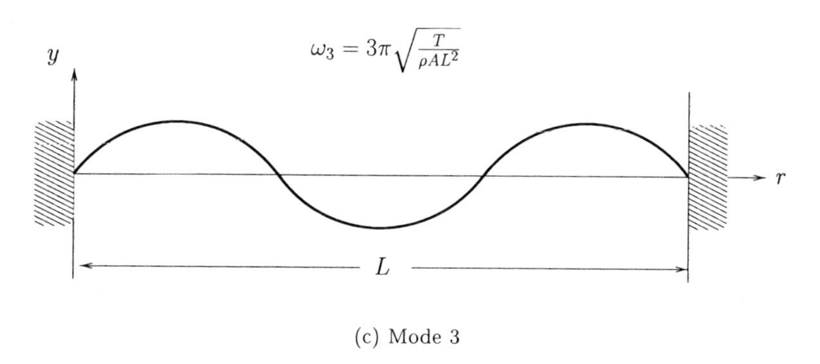

(c) Mode 3

Fig. 2.3 The first three mode shapes of the vibrating fixed uniform string

The free-body diagram of an element dr is shown in Figure 2.5. Applying Newton's second law to the force components in r direction implies

$$\left(T + \frac{\partial T}{\partial r}dr\right) - T + f(t,r)dr = \rho A(r)dr\frac{\partial^2 u(t,r)}{\partial t^2} \tag{2.29}$$

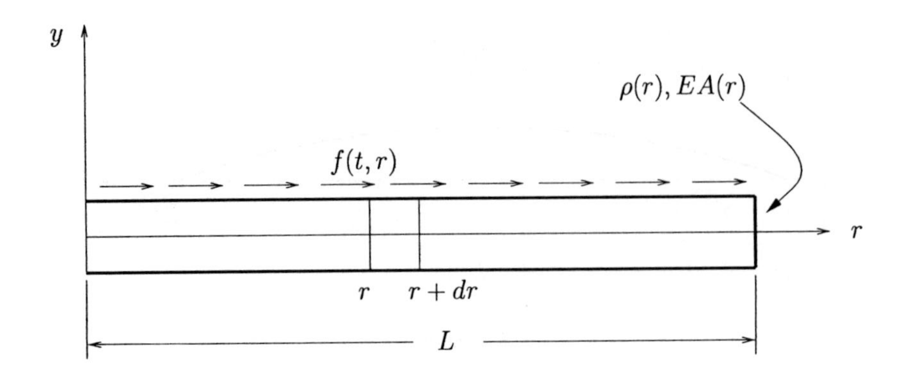

Fig. 2.4 A beam in axial vibration

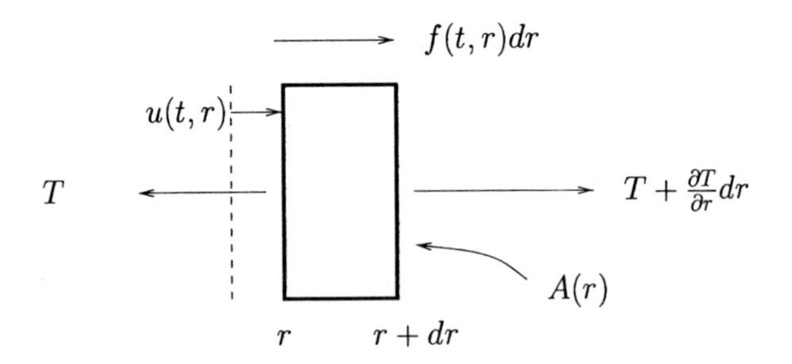

Fig. 2.5 A small element of the beam

which is equivalent to

$$\frac{\partial T}{\partial r} + f(t,r) = \rho A(r)\frac{\partial^2 u(t,r)}{\partial t^2}. \tag{2.30}$$

From Hooke's law, the tension, T, is related to the longitudinal strain by

$$T(t,r) = EA(r)\frac{\partial u(t,r)}{\partial r}, \tag{2.31}$$

where $\partial u(t,r)/\partial r$ is the longitudinal strain at point r.

Substituting (2.31) into (2.30), the PDE that governs the axial vibration of rods can be obtained [67, 16]:

$$\frac{\partial}{\partial r}\left(EA(r)\frac{\partial u(t,r)}{\partial r}\right) + f(t,r) = \rho A(r)\frac{\partial^2 u(t,r)}{\partial t^2}, \; 0 \le r \le L. \tag{2.32}$$

For a beam with uniform mass distribution and uniform cross section

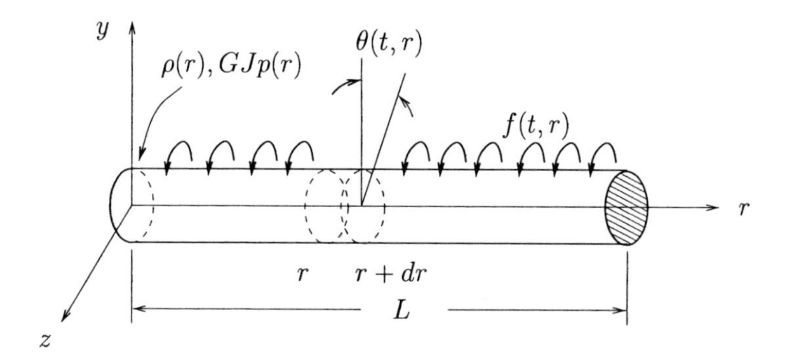

Fig. 2.6 A shaft in torsional vibration

(2.32) can be written as

$$EA\frac{\partial^2 u(t,r)}{\partial r^2} + f(t,r) = \rho A\frac{\partial^2 u(t,r)}{\partial t^2} \tag{2.33}$$

which is a partial differential equation of the type (2.1) with

$$\mathcal{L} = -EA\frac{\partial^2}{\partial r^2}$$

and

$$\mathcal{M} = \rho A.$$

2.5 Torsional vibration of shafts

Consider a shaft of length L depicted in Figure 2.6. The density, torque, distributed torque, torsional modulus of elasticity (shear modulus), polar moment of area, and angular displacement at point r are defined as $\rho(r), T(r), f(r), G(r), J_p(r)$ and $\theta(t,r)$ respectively.

The free-body diagram of an element dr is shown in Figure 2.7. Applying Newton's second law for rotary motion to torque components in the direction of r axis, we obtain

$$\left(T + \frac{\partial T}{\partial r}dr\right) - T + f(t,r)dr = \rho(r)J_p(r)dr\frac{\partial^2\theta}{\partial t^2} \tag{2.34}$$

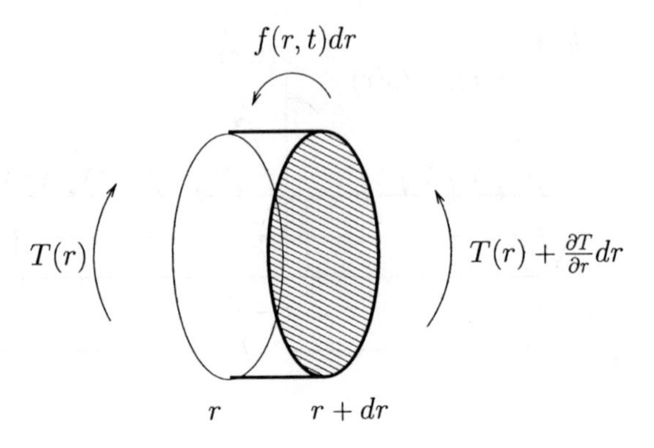

Fig. 2.7 A small element of the shaft

which is equivalent to

$$\frac{\partial T}{\partial r} + f(t,r) = \rho(r)J_p(r)\frac{\partial^2 \theta}{\partial t^2}. \tag{2.35}$$

The relationship between the torque and the angular displacement for circular cross sections can be written as [16, 29, 90, 103]:

$$T = GJ_p\frac{\partial \theta}{\partial r}, \tag{2.36}$$

where GJ_p is often called the torsional rigidity of the shaft.

Substituting (2.36) into (2.35) and re-arranging the equation gives [16]

$$\frac{\partial}{\partial r}\left(GJ_p(r)\frac{\partial \theta(t,r)}{\partial r}\right) + f(t,r) = \rho(r)J_p(r)\frac{\partial^2 \theta(t,r)}{\partial t^2},$$
$$0 \le r \le L. \tag{2.37}$$

For the case of a shaft with non-circular cross sections, the polar moment of area, J_p, must be replaced by the torsional parameter, J_t. Some examples for the torsional parameter of non-circular cross sections are given below [16, 29, 90, 4].

- **A thin closed section (a thin hollow section):**

$$J_t = \frac{4hA_s^2}{p}, \tag{2.38}$$

where A_s is the enclosed area of the hollow section, h is the section thickness, and p is the perimeter of the section.

- **A solid square section:**

$$J_t = 0.1406d^4, \tag{2.39}$$

where d is the width (or height) of the square.

- **A hollow circular section:**

$$J_t = \frac{\pi}{2}\left(r_2^4 - r_1^4\right), \tag{2.40}$$

where r_1 is the radius of the inner circle and r_2 is the radius of the outer circle.

This amounts to modifying (2.37) to

$$\frac{\partial}{\partial r}\left(GJ_t(r)\frac{\partial\theta(t,r)}{\partial r}\right) + f(t,r) = \rho(r)J_t(r)\frac{\partial^2\theta(t,r)}{\partial t^2},$$
$$0 \le r \le L, \tag{2.41}$$

where J_t is the appropriate torsional parameter. Furthermore, it can be observed that for a shaft with uniform mass distribution and fixed J_t, the partial differential equation (2.41) is of the same type as (2.1) with

$$\mathcal{L} = -GJ_t\frac{\partial^2}{\partial r^2}$$

and

$$\mathcal{M} = \rho J_t.$$

2.6 Flexural vibration of beams

Consider a thin beam of length L depicted in Figure 2.8, whose flexural rigidity at point r is $EI(r)$. Note that the flexural rigidity depends on the Young's modulus of elasticity, E, and the second moment of area, I.

Some main assumptions are:

(i) The material follows Hooke's Law.
(ii) The shear deformation is negligible compared to the bending deformation. This assumption is reasonable for thin beams.
(iii) The rotation of the element is negligible compared to the vertical translation.

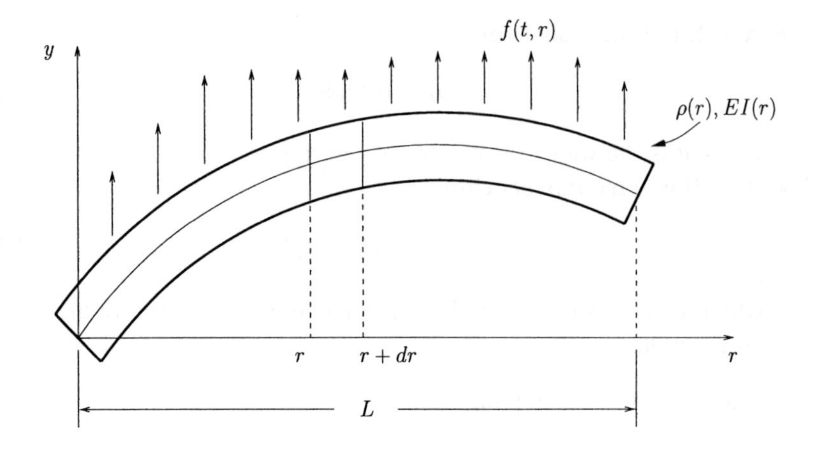

Fig. 2.8 A beam in flexural vibration

The free-body diagram of an element dr is shown in Figure 2.9, where Q denotes the shearing force and M the bending moment. Applying Newton's second law to vertical force components (y direction), we obtain

$$\left(Q(t,r) + \frac{\partial Q(t,r)}{\partial r}dr\right) - Q(t,r) + f(t,r)dr = \rho A(r)dr\frac{\partial^2 y(t,r)}{\partial t^2} \tag{2.42}$$

which is equivalent to

$$\frac{\partial Q(t,r)}{\partial r} + f(t,r) = \rho A(r)\frac{\partial^2 y(t,r)}{\partial t^2}. \tag{2.43}$$

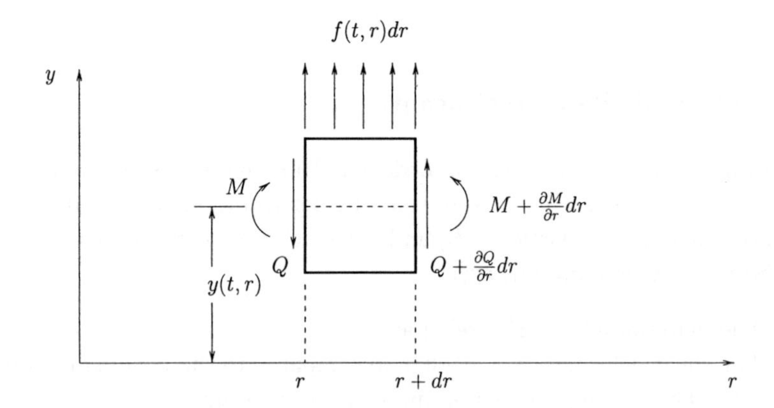

Fig. 2.9 A small element of the beam

Furthermore, considering the moment about the axis normal to r and y (out-of-page direction), we may write

$$\left(M(t,r) + \frac{\partial M(t,r)}{\partial r}dr\right) - M(t,r) + \left(Q(t,r) + \frac{\partial Q(t,r)}{\partial r}dr\right)dr + f(t,r)dr\frac{dr}{2} = 0. \tag{2.44}$$

Simplifying the above equation and canceling the higher order dr terms, the expression for the shearing force in terms of the bending moment is

$$Q(t,r) = -\frac{\partial M(t,r)}{\partial r}. \tag{2.45}$$

Substituting (2.45) into (2.43) gives

$$-\frac{\partial^2 M(t,r)}{\partial r^2} + f(t,r) = \rho A(r)\frac{\partial^2 y(t,r)}{\partial t^2}. \tag{2.46}$$

From Solid Mechanics, the bending moment can be related to the curvature of the element [67, 29, 103, 7]:

$$M(t,r) = EI(r)\frac{\partial^2 y(t,r)}{\partial r^2}. \tag{2.47}$$

Incorporating this information into (2.46) and re-arranging gives

$$\frac{\partial^2}{\partial r^2}\left(EI(r)\frac{\partial^2 y(t,r)}{\partial r^2}\right) + \rho A(r)\frac{\partial^2 y(t,r)}{\partial t^2} = f(t,r),\ 0 \le r \le L \tag{2.48}$$

which is the Bernoulli-Euler beam equation [67, 16].

Next, we describe some of the ideal boundary conditions that are commonly used [67, 16]. These "ideal" boundary conditions may not be exact for a real system. However, they could serve as good approximations of the true boundary conditions.

- **Clamped end at $r = r_o$:** The deflection and the slope of the curve are zero.

$$y(t,r_o) = 0, \qquad \left.\frac{\partial y(t,r)}{\partial r}\right|_{r=r_o} = 0. \tag{2.49}$$

- **Hinged (pinned) end at $r = r_o$:** The deflection and bending moment are zero. There is no bending moment to restrict the rotation at the end.

$$y(t, r_o) = 0, \qquad EI(r)\left.\frac{\partial^2 y(t,r)}{\partial r^2}\right|_{r=r_o} = 0. \tag{2.50}$$

- **Free end at $r = r_o$:** The shearing force and bending moment are zero since both are needed to restrict the transverse and angular displacements at the end respectively.

$$\left.\frac{\partial}{\partial r}\left(EI(r)\frac{\partial^2 y(t,r)}{\partial r^2}\right)\right|_{r=r_o} = 0, \qquad EI(r)\left.\frac{\partial^2 y(t,r)}{\partial r^2}\right|_{r=r_o} = 0. \tag{2.51}$$

The modal analysis solution for beams with simply-supported and cantilevered boundary conditions are discussed next. The solution for beams with some other boundary conditions can be obtained in a similar way.

2.6.1 *A simply-supported beam*

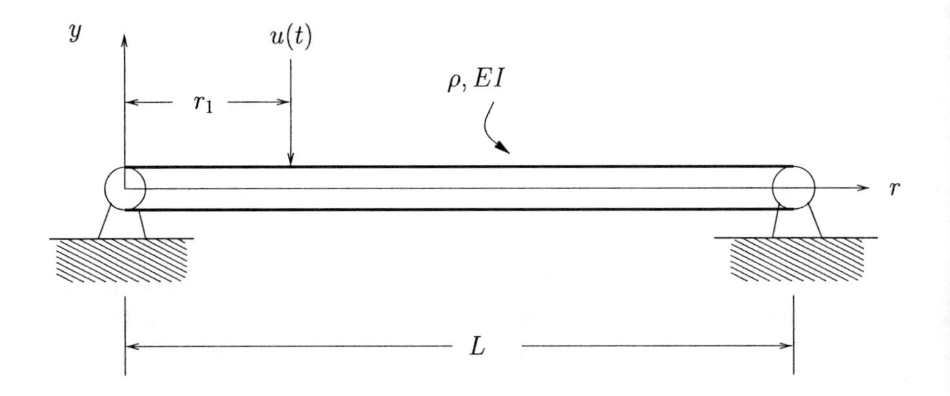

Fig. 2.10 A simply-supported beam

Obtaining an analytic model for a beam with simply-supported boundary conditions (pinned at both ends) is quite straightforward. Consider a simply-supported uniform beam where a point force $u(t)$ is acting at a point $r = r_1$ as depicted in Figure 2.10. The partial differential equation that governs the dynamics of the beam is is given in (2.48). Corresponding to this partial differential equation are the boundary conditions:

$$\begin{aligned} y(t,0) &= 0, \\ y(t,L) &= 0, \\ EI\left.\frac{\partial^2 y(t,r)}{\partial r^2}\right|_{r=0} &= 0 \text{ and} \\ EI\left.\frac{\partial^2 y(t,r)}{\partial r^2}\right|_{r=L} &= 0. \end{aligned} \tag{2.52}$$

Notice that according to the terminology of Section 2.2

$$\begin{aligned} \mathcal{L} &= EI\frac{\partial^4}{\partial r^4}, \\ \mathcal{M} &= \rho A, \\ \mathcal{B}_1 &= 1, \\ \mathcal{B}_2 &= EI\frac{\partial^2}{\partial r^2}, \\ f(t,r) &= u(t)\delta(r-r_1) \end{aligned} \tag{2.53}$$

and $\mathcal{R} = [0, L]$. The orthogonality conditions can be written as

$$\int_0^L \phi_i(r)\phi_j(r)\rho A dr = \delta_{ij}, \tag{2.54}$$

$$\int_0^L \phi_i(r)\phi_j^{''''}(r)EI dr = \omega_i^2\delta_{ij} \tag{2.55}$$

where ω_is are the solutions to the following eigenvalue problem:

$$\phi_i^{''''}(r) - \lambda_i^4\phi_i(r) = 0, \tag{2.56}$$

where, by definition,

$$\lambda_i^4 = \frac{\rho A\omega_i^2}{EI}. \tag{2.57}$$

The eigenfunction, $\phi_i(r)$, has to satisfy the corresponding boundary conditions:

$$\begin{aligned} \phi_i(0) &= 0, \\ \phi_i(L) &= 0, \\ \phi_i^{''}(0) &= 0 \text{ and} \\ \phi_i^{''}(L) &= 0. \end{aligned} \tag{2.58}$$

The general solution to the eigenvalue problem is of the form [67, 16]:

$$\phi_i(r) = A_i \sin \lambda_i r + B_i \cos \lambda_i r + C_i \sinh \lambda_i r + D_i \cosh \lambda_i r. \quad (2.59)$$

The coefficients A_i, B_i, C_i and D_i are to be determined from the boundary conditions. The first and third boundary conditions in (2.58) imply

$$B_i = D_i = 0.$$

The remainder of the boundary conditions, on the other hand, imply that

$$C_i = 0$$

and

$$\sin \lambda_i L = 0.$$

Thus, λ_is are found to be

$$\lambda_i = \frac{i\pi}{L}, \qquad i = 1, 2, \dots \; . \quad (2.60)$$

Therefore, $\phi_i(r)$ can be written as

$$\phi_i(r) = A_i \sin \lambda_i r.$$

Substituting this expression into the orthogonality condition (2.55), the expression for A_i can be found as

$$A_i = \sqrt{\frac{2}{\rho A L}}.$$

Hence, to summarize, the mode shapes for the simply-supported beam are given by sinusoidal functions:

$$\phi_i(r) = \sqrt{\frac{2}{\rho A L}} \sin\left(\frac{i\pi r}{L}\right), \; i = 1, 2, \dots \quad (2.61)$$

and the corresponding natural frequencies are

$$\omega_i = \left(\frac{i\pi}{L}\right)^2 \sqrt{\frac{EI}{\rho A}}, \; i = 1, 2, \dots \; . \quad (2.62)$$

The transfer function between the applied force $\hat{u}(s)$ and the transverse deflection of the beam $\hat{y}(s, r)$ is found to be

$$\frac{\hat{y}(s,r)}{\hat{u}(s)} = \sum_{i=1}^{\infty} \frac{\phi_i(r_1)\phi_i(r)}{s^2 + \omega_i^2}. \tag{2.63}$$

2.6.2 *A cantilevered beam*

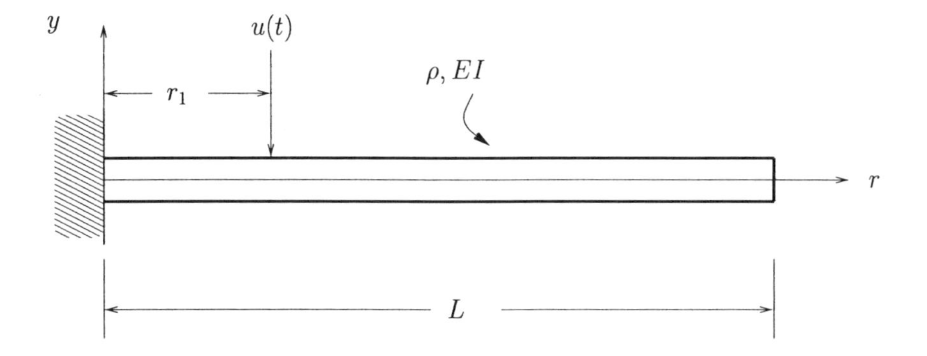

Fig. 2.11 A cantilevered beam

Consider a cantilevered uniform beam shown in Figure 2.11. The boundary conditions are:

$$\begin{aligned} y(t,0) &= 0, \\ \left.\frac{\partial y(t,r)}{\partial r}\right|_{r=0} &= 0, \\ \left.\frac{\partial^2 y(t,r)}{\partial r^2}\right|_{r=L} &= 0 \text{ and} \\ \left.\frac{\partial^3 y(t,r)}{\partial r^3}\right|_{r=L} &= 0. \end{aligned} \tag{2.64}$$

Once again we assume that a point force $u(t)$ is acting at point $r = r_1$.

The general solution to the eigenvalue problem is given in (2.59) and the coefficients can be found from the boundary conditions:

$$\begin{aligned} \phi_i(0) &= 0, \\ \phi_i^{'}(0) &= 0, \\ \phi_i^{''}(L) &= 0 \quad \text{and} \\ \phi_i^{'''}(L) &= 0. \end{aligned} \tag{2.65}$$

The first boundary condition leads to $B_i + D_i = 0$, while the second requires $A_i + C_i = 0$. The solution, then, is of the form:

$$\phi_i(r) = A_i\left(\sin\lambda_i r - \sinh\lambda_i r\right) + B_i\left(\cos\lambda_i r - \cosh\lambda_i r\right). \tag{2.66}$$

The last two boundary conditions yield two equations:

$$A_i\left(\sin\lambda_i L + \sinh\lambda_i L\right) + B_i\left(\cos\lambda_i L + \cosh\lambda_i L\right) = 0 \quad \text{and} \tag{2.67}$$

$$A_i\left(\cos\lambda_i L + \cosh\lambda_i L\right) - B_i\left(\sin\lambda_i L - \sinh\lambda_i L\right) = 0. \tag{2.68}$$

Both equations are solved in terms of A_i. Knowing that

$$\cosh\lambda_i L^2 - \sinh\lambda_i L^2 = 1$$

and

$$\cos\lambda_i L^2 + \sin\lambda_i L^2 = 1$$

along with the fact that $A_i \neq 0$ for non-trivial solution, the following relationship is obtained:

$$\cos\lambda_i L \cosh\lambda_i L = -1. \tag{2.69}$$

This transcendental equation needs to be solved for $\lambda_i L$ either graphically or numerically. Since λ_i can be related to ω_i by (2.57), i.e.

$$\omega_i = \lambda_i^2\sqrt{\frac{EI}{\rho A}}, \tag{2.70}$$

the solution to (2.69) yields the natural frequencies of the cantilevered beam. There is an infinite number of solutions to this equation, so there are an infinite set of natural frequencies. The first three solutions of (2.69) are:

$$\begin{aligned} \lambda_1 &= 1.875/L, \\ \lambda_2 &= 4.694/L \quad \text{and} \\ \lambda_3 &= 7.855/L. \end{aligned} \tag{2.71}$$

The normalized mode shapes, i.e. $A_i = 1$, are as follows [67, 16]:

$$\phi_i(r) = (\sin \lambda_i r - \sinh \lambda_i r) - \alpha_i (\cos \lambda_i r - \cosh \lambda_i r) \tag{2.72}$$

where

$$\alpha_i = \frac{(\sin \lambda_i L + \sinh \lambda_i L)}{(\cos \lambda_i L + \cosh \lambda_i L)}.$$

The transfer function between the applied force $\hat{u}(s)$ and the transverse deflection of the beam $\hat{y}(s, r)$ has a similar form as in (2.63), with the mode shapes and natural frequencies as in (2.72) and (2.70) respectively. It is important to note that although there is theoretically an infinite number of modes, a real system has only a finite but very large number of modes. The first three mode shapes of a vibrating cantilevered beam are shown in Figure 2.12.

2.7 Transverse vibration of thin plates

This section is concerned with modeling of thin plates. The main assumptions are:

(i) The plate has a uniform thickness.
(ii) The shear deformation, stress in lateral direction and rotational inertia of the plate are ignored.

A thin plate with dimensions of $a \times b \times h$ is shown in Figure 2.13. The Young's modulus of elasticity, density and Poisson's ratio of the plate are denoted by E, ρ and ν respectively. Consider the plate element in Figure 2.14. The origin is located on the mid-plane of the plate, i.e. the plate's neutral surface, where it experiences no longitudinal bending strain.

The deflection in X and Y directions can be expressed as follows [46, 63, 111]:

$$\begin{aligned} u &= -z \frac{\partial w}{\partial x} \\ v &= -z \frac{\partial w}{\partial y}, \end{aligned} \tag{2.73}$$

where w is the deflection in Z direction.

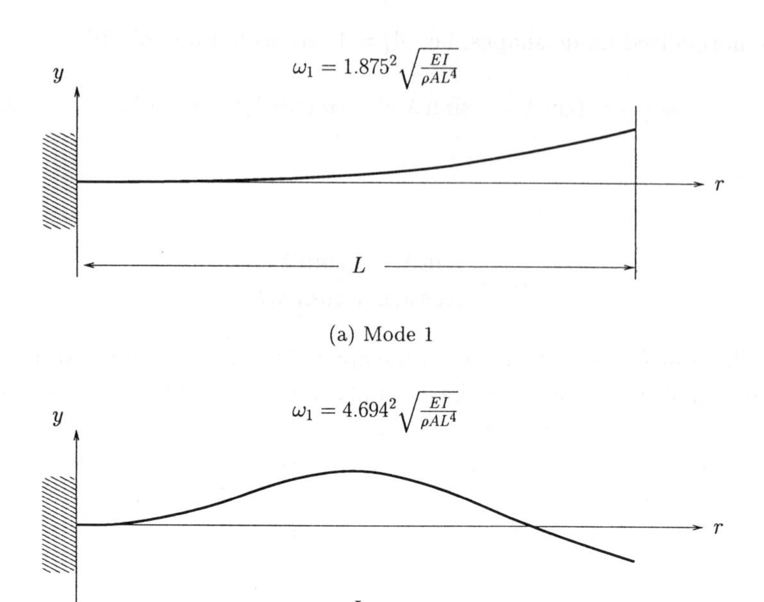

(a) Mode 1

(b) Mode 2

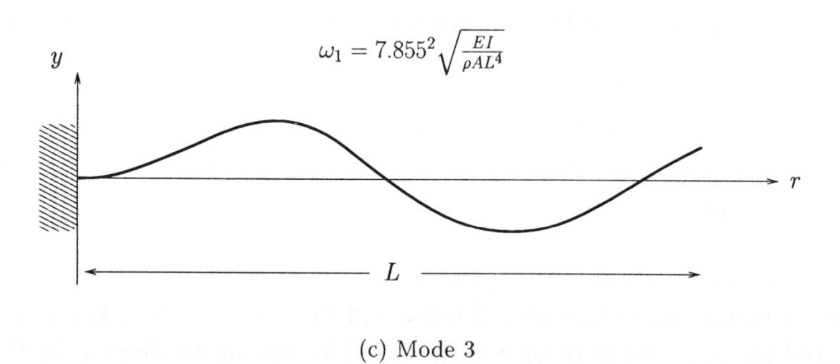

(c) Mode 3

Fig. 2.12 The first three mode shapes of the vibrating fixed uniform beam

Using Hooke's Law [46, 63, 111], the expression for the strains can be obtained from (2.73):

$$\begin{aligned}\epsilon_x &= \frac{\partial u}{\partial x} = -z\frac{\partial^2 w}{\partial x^2},\\ \epsilon_y &= \frac{\partial v}{\partial y} = -z\frac{\partial^2 w}{\partial y^2} \quad \text{and}\\ \gamma_{xy} &= \frac{\partial u}{\partial y} + \frac{\partial v}{\partial x} = -2z\frac{\partial^2 w}{\partial x \partial y},\end{aligned} \tag{2.74}$$

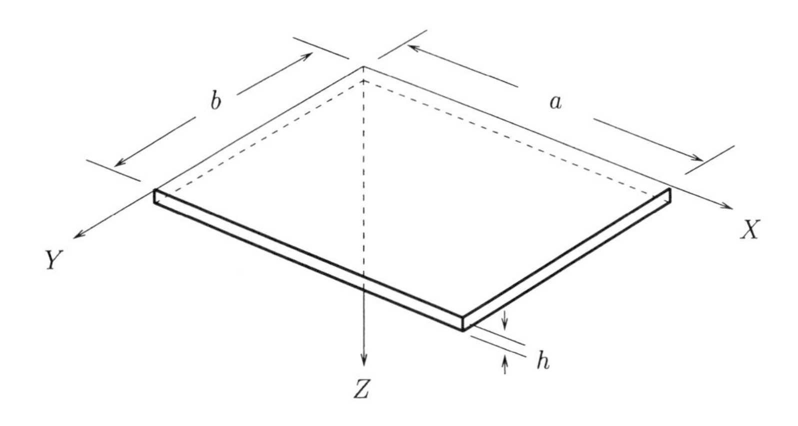

Fig. 2.13 A thin plate in transverse vibration

where γ_{xy} is the shear strain, while ϵ_x and ϵ_y are the longitudinal strains in X and Y directions respectively.

From Hooke's Law, strains are related to stresses as follows [46, 63, 111]:

$$\begin{aligned} \epsilon_x &= \frac{1}{E}(\sigma_x - \nu\sigma_y), \\ \epsilon_y &= \frac{1}{E}(\sigma_y - \nu\sigma_x) \quad \text{and} \\ \gamma_{xy} &= \frac{1}{G}\tau_{xy} = 2\frac{(1+\nu)}{E}\tau_{xy}, \end{aligned} \tag{2.75}$$

where G is the shear modulus and can be related to Young's modulus as described above. Further, τ_{xy} is the shear stress, while γ_x and γ_y are the longitudinal stresses in X and Y directions respectively. Hence, the stresses can be determined from (2.74) and (2.75):

$$\begin{aligned} \tau_{xy} &= -\frac{Ez}{(1+\nu)}\frac{\partial^2 w}{\partial x \partial y}, \\ \sigma_x &= -\frac{Ez}{(1-\nu^2)}\left(\frac{\partial^2 w}{\partial x^2} + \nu\frac{\partial^2 w}{\partial y^2}\right) \quad \text{and} \\ \sigma_y &= -\frac{Ez}{(1-\nu^2)}\left(\frac{\partial^2 w}{\partial y^2} + \nu\frac{\partial^2 w}{\partial x^2}\right). \end{aligned} \tag{2.76}$$

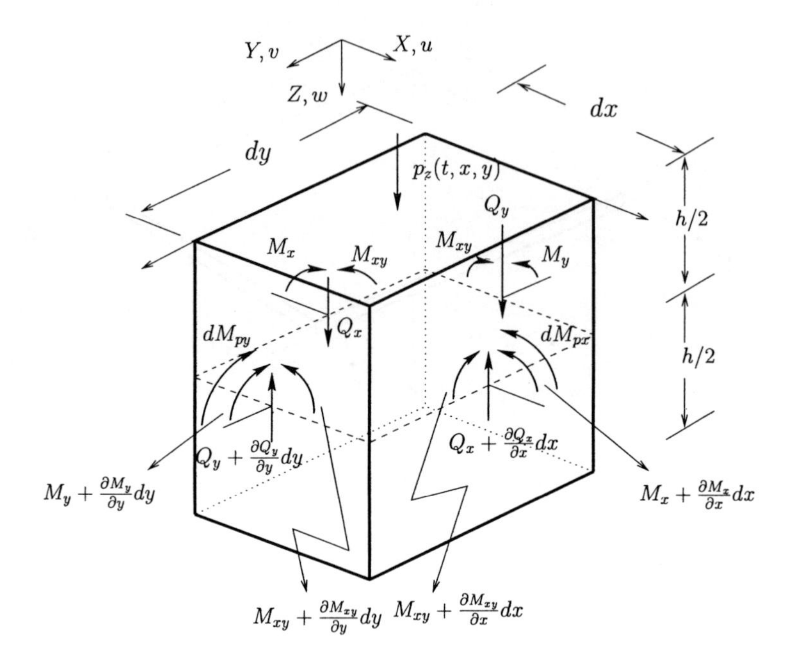

Fig. 2.14 A small element of the plate

The moment per unit length M_x is calculated by integrating the corresponding stress across the plate thickness,

$$M_x = \int_{-\frac{h}{2}}^{\frac{h}{2}} z\sigma_x dz = -D\left(\frac{\partial^2 w}{\partial x^2} + \nu\frac{\partial^2 w}{\partial y^2}\right) \tag{2.77}$$

where D is the *flexural rigidity* of the plate [99, 111]:

$$D = \frac{Eh^3}{12(1-\nu^2)}. \tag{2.78}$$

Similarly for M_y,

$$M_y = \int_{-\frac{h}{2}}^{\frac{h}{2}} z\sigma_y dz = -D\left(\frac{\partial^2 w}{\partial y^2} + \nu\frac{\partial^2 w}{\partial x^2}\right). \tag{2.79}$$

The torsional moment per unit length M_{xy},

$$M_{xy} = \int_{-\frac{h}{2}}^{\frac{h}{2}} z\tau_{xy} dz = -D(1-\nu)\frac{\partial^2 w}{\partial x \partial y}. \tag{2.80}$$

The external moments per unit length are denoted by dM_{px} and dM_{py}, while p_z is the pressure in Z direction. Q_x and Q_y denote the shearing forces per unit length as shown in Figure 2.14. Applying Newton's second law to vertical forces (in Z direction), we obtain

$$Q_x dy - \left(Q_x + \frac{\partial Q_x}{\partial x}dx\right)dy + Q_y dx - \left(Q_y + \frac{\partial Q_y}{\partial y}dy\right)dx + p_z(t,x,y)\,dx\,dy - \rho(x,y)\,h\,dx\,dy\frac{\partial^2 w}{\partial t^2} = 0. \tag{2.81}$$

Dividing by $dx\,dy$ yields

$$-\frac{\partial Q_x}{\partial x} - \frac{\partial Q_y}{\partial y} + p_z(t,x,y) - \rho(x,y)h\frac{\partial^2 w}{\partial t^2} = 0. \tag{2.82}$$

Taking moment equilibrium about the X axis and ignoring the rotational inertia of the plate, we may write

$$p_z(t,x,y)\,dxdy\frac{dy}{2} - \frac{\partial Q_x}{\partial x}dxdy\frac{dy}{2} - \left(Q_y + \frac{\partial Q_y}{\partial y}dy\right)dxdy - \frac{\partial M_y}{\partial y}dydx - \frac{\partial M_{xy}}{\partial x}dxdy - dM_{py}dx = 0. \tag{2.83}$$

Dividing by $dx\,dy$ and ignoring the higher order term dy, we obtain

$$-Q_y = \frac{\partial M_y}{\partial y} + \frac{\partial M_{xy}}{\partial x} + \frac{\partial M_{py}}{\partial y}. \tag{2.84}$$

Similarly, the moment equilibrium can be taken about the Y axis to obtain

$$-Q_x = \frac{\partial M_x}{\partial x} + \frac{\partial M_{xy}}{\partial y} + \frac{\partial M_{px}}{\partial x}. \tag{2.85}$$

In this case $M_{xy} = M_{yx}$ is assumed due to complementary shear stress, $\tau_{xy} = \tau_{yx}$. Differentiating (2.84) with respect to y,

$$-\frac{\partial Q_y}{\partial y} = \frac{\partial^2 M_y}{\partial y^2} + \frac{\partial^2 M_{xy}}{\partial x \partial y} + \frac{\partial^2 M_{py}}{\partial y^2}. \tag{2.86}$$

Differentiating (2.85) with respect to x,

$$-\frac{\partial Q_x}{\partial x} = \frac{\partial^2 M_x}{\partial x^2} + \frac{\partial^2 M_{xy}}{\partial x \partial y} + \frac{\partial^2 M_{px}}{\partial x^2}. \tag{2.87}$$

Substituting (2.86) and (2.87) into the vertical equilibrium equation (2.82) gives

$$\left(\frac{\partial^2 M_x}{\partial x^2} + 2\frac{\partial^2 M_{xy}}{\partial x \partial y} + \frac{\partial^2 M_y}{\partial y^2}\right) + \left(\frac{\partial^2 M_{px}}{\partial x^2} + \frac{\partial^2 M_{py}}{\partial y^2}\right)$$
$$+p_z(t,x,y) - \rho(x,y)h\frac{\partial^2 w}{\partial t^2} = 0. \tag{2.88}$$

Now, the moment expressions in (2.77), (2.79) and (2.80) can be differentiated twice to obtain $\partial^2 M_x/\partial x^2, \partial^2 M_y/\partial y^2$ and $\partial^2 M_{xy}/\partial x \partial y$. Substituting these into (2.88) and simplifying yields the PDE of a thin plate under transverse vibration:

$$\rho(x,y)h\frac{\partial^2 w}{\partial t^2} + D\nabla^4 w(t,x,y) = \frac{\partial^2 M_{px}}{\partial x^2} + \frac{\partial^2 M_{py}}{\partial y^2} + p_z(t,x,y), \tag{2.89}$$

where

$$\nabla^4 w = \frac{\partial^4 w}{\partial x^4} + 2\frac{\partial^4 w}{\partial x^2 \partial y^2} + \frac{\partial^4 w}{\partial y^4}. \tag{2.90}$$

2.7.1 *Simply-supported rectangular thin plates*

Consider a rectangular uniform thin plate whose edges are pinned (hinged). The boundary conditions are:

$$w(t,x,y) = 0, M_x = -D\left(\frac{\partial^2 w}{\partial x^2} + \nu\frac{\partial^2 w}{\partial y^2}\right) = 0, \forall x = 0, a; 0 \le y \le b \text{ and}$$
$$w(t,x,y) = 0, M_y = -D\left(\frac{\partial^2 w}{\partial y^2} + \nu\frac{\partial^2 w}{\partial x^2}\right) = 0, \forall y = 0, b; 0 \le x \le a. \tag{2.91}$$

Natural frequencies and mode shapes can be determined in a similar manner to what we did for flexural vibration of beams. Hence, the procedure is not repeated again. The key difference is that each mode shape is now a function of two spatial variables, x and y. It can be shown that the normalized mode shape of mode (m,n), where m and n are the mode numbers in X and Y directions respectively, is

$$\phi_{mn}(x,y) = \frac{2}{\sqrt{ab\rho h}} \sin\frac{m\pi x}{a} \sin\frac{n\pi y}{b}, \tag{2.92}$$

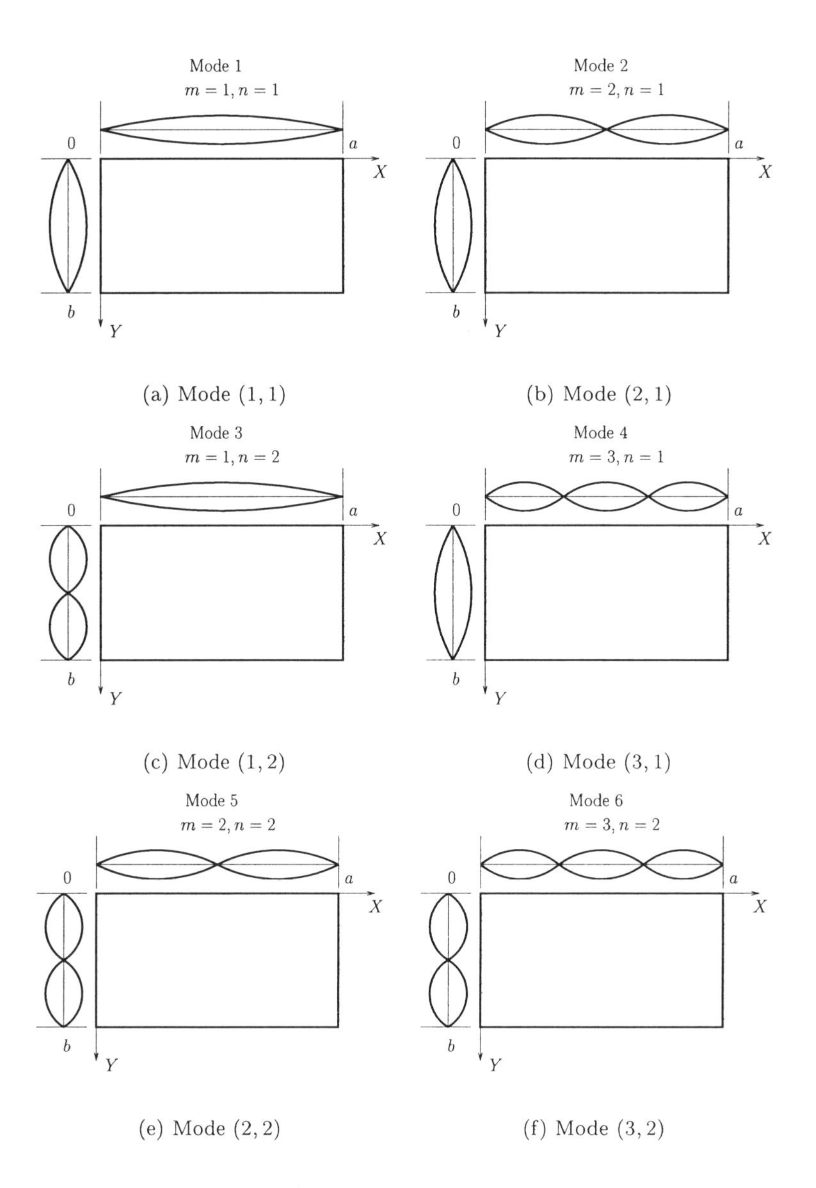

Fig. 2.15 Mode shapes of transverse vibration of a simply-supported plate

where the natural frequency of mode (m, n) is

$$\omega_{mn} = \pi^2 \sqrt{\frac{D}{\rho h}} \left(\frac{m^2}{a^2} + \frac{n^2}{b^2} \right). \tag{2.93}$$

Examples of typical simply-supported mode shapes are shown in Figure 2.15.

2.7.2 Plates with more general boundary conditions

Modal analysis can be used modal to obtain exact solutions to the partial differential equations that describe dynamics of plates with specific boundary conditions, e.g. a plate with simply-supported boundary conditions. However, exact solutions are usually very difficult to obtain for plates with more general boundary conditions. This is due to the fact that the system eigenfunctions have to simultaneously satisfy the associated eigenvalue problem and boundary conditions. Such eigenfunctions might not exist, hence we have to rely on approximate solutions when dealing with such systems. A particular way to solve for such systems is to use the Ritz method (also sometimes called the Rayleigh-Ritz method). In this method, the eigenfunction associated with each mode is approximated by a linear combination of n admissible functions. These functions only need to satisfy the geometric boundary conditions of the system. Hence, the solution is estimated by

$$w(t,x,y) = \phi(x,y)'\, q(t), \tag{2.94}$$

where $\phi \in \mathbf{R}^n$ contains the admissible functions and $q \in \mathbf{R}^n$ is a time-dependent vector.

We obtain the equation of motion for the system from its kinetic and potential energies. The kinetic energy is obtained by incorporating the solution in (2.94):

$$\begin{aligned} \mathcal{T} &= \frac{1}{2}\int_0^b \int_0^a \rho(x,y)\, h \left(\frac{\partial w}{\partial t}\right)^2 dx\, dy, \\ &= \frac{1}{2}\, \dot{q}(t)'\, M\, \dot{q}(t) \end{aligned} \tag{2.95}$$

where M is the system's mass matrix:

$$M = \int_0^b \int_0^a \rho(x,y)\, h\, \phi\, \phi' dx\, dy.$$

We consider the bending strain energy of the plate [101] as the system's potential energy:

$$\mathcal{V} = \frac{1}{2}\int_0^b \int_0^a D \left\{ \left(\frac{\partial^2 w}{\partial x^2} + \frac{\partial^2 w}{\partial y^2} \right)^2 \right.$$
$$\left. -2(1-\nu) \left[\frac{\partial^2 w}{\partial x^2}\frac{\partial^2 w}{\partial y^2} - \left(\frac{\partial^2 w}{\partial x \partial y} \right)^2 \right] \right\} dx\, dy. \tag{2.96}$$

Equation (2.96) can be written as

$$\mathcal{V} = \frac{1}{2} q(t)'\, K\, q(t), \tag{2.97}$$

where K is the system's stiffness matrix:

$$K = \int_0^b \int_0^a D \begin{bmatrix} \frac{\partial^2 \phi}{\partial x^2} & \frac{\partial^2 \phi}{\partial y^2} & \frac{\partial^2 \phi}{\partial x \partial y} \end{bmatrix} \times \begin{bmatrix} 1 & \nu & 0 \\ \nu & 1 & 0 \\ 0 & 0 & 2(1-\nu) \end{bmatrix}$$
$$\times \begin{bmatrix} \frac{\partial^2 \phi'}{\partial x^2} \\ \frac{\partial^2 \phi'}{\partial y^2} \\ \frac{\partial^2 \phi'}{\partial x \partial y} \end{bmatrix} dx\, dy. \tag{2.98}$$

The effect of external forces/moments can be obtained by calculating the virtual work done by them:

$$\delta W = \int_0^b \int_0^a \left(\frac{\partial^2 M_{px}}{\partial x^2} + \frac{\partial^2 M_{py}}{\partial y^2} + p_z \right) \delta w\, dx\, dy$$
$$= F'\, \delta q, \tag{2.99}$$

where

$$F = \int_0^b \int_0^a \left(\frac{\partial^2 M_{px}}{\partial x^2} + \frac{\partial^2 M_{py}}{\partial y^2} + p_z \right) \phi\, dx\, dy.$$

The equation of motion of the system is:

$$M\, \ddot{q} + K\, \dot{q} = F. \tag{2.100}$$

Solving the eigenvalue problem of (2.100) for $F = 0$, we can obtain the normalized eigenvectors $\Phi = [\phi_1 \dots \phi_n]$ and eigenvalues $[\omega_1^2 \dots \omega_n^2]$ of the plate. The following orthogonality properties can be obtained:

$$\begin{aligned} \Phi' M \Phi &= I \\ \Phi' K \Phi &= \Lambda, \end{aligned} \tag{2.101}$$

where I is a unit matrix and

$$\Lambda = \text{diag}(\omega_1^2 \dots \omega_n^2).$$

2.8 Modeling of piezoelectric laminate beams

Piezoelectricity was first discovered in 1880 by Pierre and Paul-Jacques Curie. They observed that various crystals, such as tourmaline, Rouchelle salt and quartz, generated electrical charges on their surface when they were mechanically strained in certain directions [8]. In the following year, they discovered the converse effect, i.e. applying an electric field to the crystals change their shape.

The existence of the piezoelectric effect in some materials implies that they have the potential to be employed as actuators and sensors. Suppose a piezoelectric material is bonded to a structure. A mechanical deformation can be induced in the structure by applying a voltage to the piezoelectric material, so employing the piezoelectric element as an actuator. On the other hand, a structural deformation can be sensed by the piezoelectric element by measuring the amount of electrical charge the material produces. Unfortunately, the piezoelectric effect in natural crystals is rather weak so they cannot be used effectively as actuators.

Recent development in the field of materials science has provided piezoelectric materials that have sufficient coupling between electrical and mechanical energies. Some of the commonly used piezoelectric materials are poly-vinylidene fluoride (PVDF), a semi-crystalline polymer film, and lead zirconate titanate (PZT), a piezoelectric ceramic material. For a more detailed discussion of the electromechanical properties of piezoelectric materials, the reader is referred to [83, 80, 15, 56, 25].

Almost all of the experiments presented in this book are carried out on piezoelectric laminate structures. Therefore, this section is devoted to modeling of flexural vibration of a piezoelectric laminate beam, which is our main experimental test-bed.

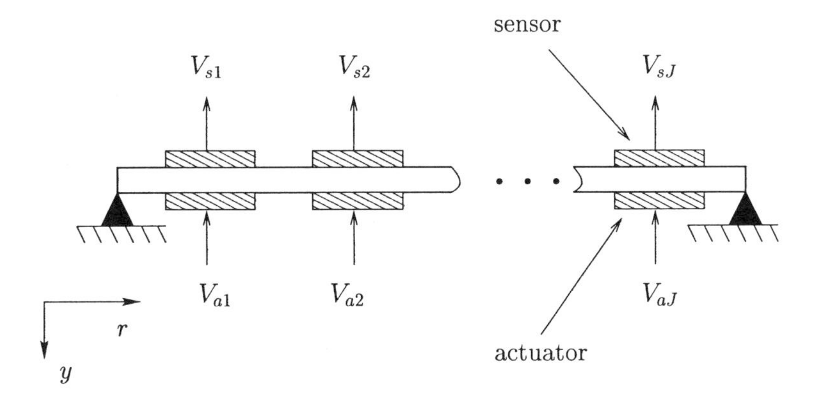

Fig. 2.16 A beam with collocated piezoelectric actuators and sensors

2.8.1 *Dynamics of a piezoelectric laminate beam*

Consider a homogeneous Euler-Bernoulli beam with a number of piezoelectric actuator-sensor pairs attached to it (a piezoelectric laminate beam) as shown in Figure 2.16. Suppose there are J collocated actuator-sensor pairs distributed over the structure. Piezoelectric patches on one side of the beam act as sensors, while those on the other side serve as actuators. Suppose that the jth piezoelectric actuator and sensor have dimensions of $L_{pj} \times W_{pj} \times h_{pj}$, where h_{pj} is the thickness of each patch, while the beam has dimensions of $L \times W \times h$ (see Figure 2.17). The applied voltages to the actuating patches are denoted by

$$V_a(t) = [V_{a1}(t), \ldots, V_{aJ}(t)]'.$$

The partial differential equation which governs the dynamics of the homogeneous beam is as follows [12, 67],

$$EI\,\frac{\partial^4 y(t,r)}{\partial r^4} + \rho A \frac{\partial^2 y(t,r)}{\partial t^2} = \frac{\partial^2 M_{pr}(t,r)}{\partial r^2}, \tag{2.102}$$

where all parameters are defined in previous sections. In this particular case, the external disturbance on the beam is assumed to be in the form of a bending moment, M_{pr}, since only the flexural vibration case is considered.

We need to derive the bending moments exerted on the beam by the piezoelectric actuators. The actuators generate bending moments on the beam when voltages are applied to them. It is assumed that the mass and stiffness contributions of the patches are much smaller than those of the

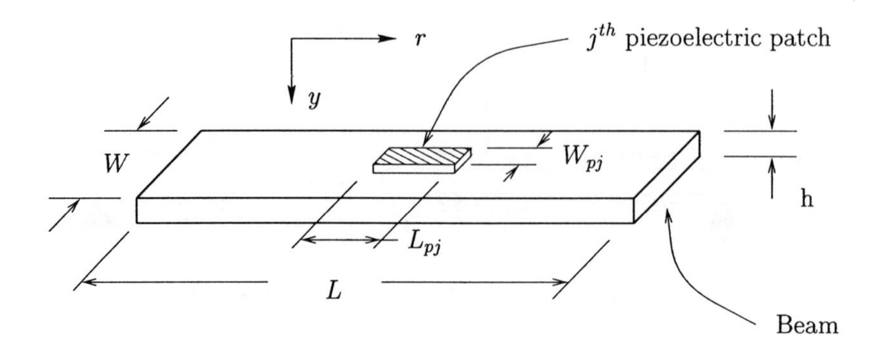

Fig. 2.17 A beam with the jth piezoelectric patch attached to it

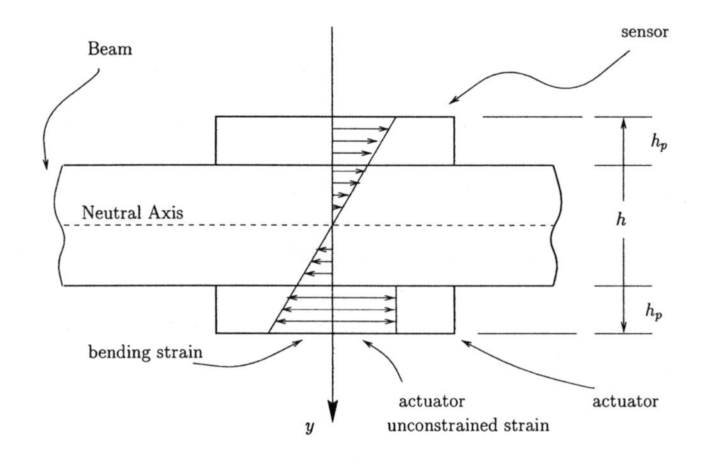

Fig. 2.18 Strain distribution of a beam section

beam so they can be neglected. This simplifies the solution for the beam dynamics since a uniform beam can be considered.

The approach presented below follows the procedures in [18] and [25]. Consider the jth piezoelectric actuator attached to the beam. Let us drop the subscript j for the time being, realizing that the properties that are discussed in the following, belong to the jth actuator. The overall longitudinal strain inside the actuator is contributed by the induced longitudinal strain due to bending ϵ_r and the unconstrained strain ϵ_p. The unconstrained strain is the strain produced by the piezoelectric patch when the patch is free to expand or contract [18, 25], i.e. the strain due to the applied voltage. An expression for the unconstrained strain (see Figure 2.18) of the piezoelectric material due to the applied voltage is [25, 3]

$$\epsilon_p = \left(\frac{d_{31}}{h_p} \right) V_{aj}(t). \tag{2.103}$$

The piezoelectric charge constant and the applied voltage to the jth actuator are denoted by d_{31} and $V_{aj}(t)$ respectively.

Using Hooke's Law to obtain an expression for stress in terms of strain, we may write

$$\sigma_{pr} = E_p \left(\epsilon_r - \epsilon_p \right) \tag{2.104}$$

$$\sigma_r = E \epsilon_r, \tag{2.105}$$

where σ_{pr} and σ_r are the longitudinal stresses of the actuator and the beam respectively, in r direction, and E_p is the Young's modulus of elasticity of the actuator.

The strain distribution across the beam thickness can be considered to be linear for pure flexural case, $\epsilon_r = \alpha\, y$, as shown in Figure 2.18. The strain gradient α is determined from the moment equilibrium equation about the neutral axis of the beam:

$$\int_{-\frac{h}{2}}^{\frac{h}{2}} y\, \sigma_r\, dy + \int_{\frac{h}{2}}^{\frac{h}{2}+h_p} y\, \sigma_{pr}\, dy = 0. \tag{2.106}$$

It is assumed that the piezoelectric patch is bonded to the beam perfectly. This amounts to having no shearing effect in the patch-structure interface. It is also assumed that the patches are relatively much thinner than the thickness of the beam, so the position of the neutral axis does not change.

From (2.106), the expression for the relationship between the unconstrained actuator strain ϵ_p and the bending strain gradient α is found to be

$$\alpha = \kappa\, \varepsilon_p, \tag{2.107}$$

where

$$\kappa = \frac{12\, E_p\, h_p\, (h_p + h)}{2\, E\, h^3 + E_p\, \left[(h + 2\, h_p)^3 - h^3 \right]}. \tag{2.108}$$

Suppose that the ends of the jth piezoelectric patch are located at r_{1j} and r_{2j} along the r axis. The bending moment experienced by the beam M_{prj} can be determined from the first integral term in (2.106) with the use of (2.107):

$$M_{prj} = K_j \left[H(r - r_{1j}) - H(r - r_{2j}) \right] V_{aj}(t), \tag{2.109}$$

where

$$K_j = \frac{\kappa_j E d_{31j} h^3 W_{pj}}{12 h_{pj}} \tag{2.110}$$

which depends on the properties of the beam and the jth piezoelectric patch. Note that $H(\cdot)$ is a step function that is:

$$H(r - r_{1j}) = \begin{cases} 0 & r < r_{1j} \\ 1 & r \geq r_{1j}. \end{cases}$$

The forcing term in the partial differential equation (2.102) can then be determined from the M_{prj} expression, using the property of Dirac delta function [52]:

$$\int_{-\infty}^{\infty} \delta^{(n)}(t - \theta)\, \phi(t)\, dt = (-1)^n \, \phi^{(n)}(\theta), \tag{2.111}$$

where $\delta^{(n)}$ is the nth derivative of δ and ϕ is continuous at θ.

Having obtained the piezoelectric moment expression, we need to solve the PDE (2.102). The modal analysis technique is used for this purpose by considering a typical solution of the form

$$y(t, r) = \sum_{i=1}^{\infty} \phi_i(r)\, q_i(t). \tag{2.112}$$

The eigenfunction $\phi_i(r)$ satisfies orthogonality conditions stated in Section 2.6.

A set of decoupled second order ordinary differential equations can be obtained from the partial differential equation of the beam (2.102) by using the orthogonality properties and Dirac delta function property (2.111) as well as the solution in (2.112). Considering the bending moments generated by all J piezoelectric actuators, the following ODEs are obtained:

$$\ddot{q}_i(t) + 2\zeta_i \omega_i \dot{q}_i(t) + \omega_i^2 q_i(t) = \frac{1}{\rho A} \sum_{j=1}^{J} K_j \Psi_{ij} V_{aj}(t), \tag{2.113}$$

where $i = 1, 2, \cdots$, and $q_i(t)$ is the generalized coordinate of mode i and the subscript j denotes the jth actuator. A proportional damping term ζ_i for each mode i has been added into the equation of motion in (2.113).

Further, Ψ_{ij} can be obtained from

$$\Psi_{ij} = \int_0^L \phi_i(r) \left[\frac{d\delta(r - r_{1j})}{dr} - \frac{d\delta(r - r_{2j})}{dr} \right] dr. \tag{2.114}$$

Applying the Laplace Transform to (2.113), assuming zero initial conditions, the MIIO (Multiple-Input, Infinite-Output) transfer function from the applied actuator-voltages, $V_a(s) = [V_{a1}(s), \ldots, V_{aJ}(s)]'$, to the beam deflection $y(s, r)$ is found to be

$$G(s,r) = \sum_{i=1}^{\infty} \frac{\phi_i(r)\, P_i}{s^2 + 2\zeta_i\, \omega_i\, s + \omega_i^2}, \tag{2.115}$$

where

$$P_i = \frac{1}{\rho A}[K_1 \Psi_{i1}, \ldots, K_J \Psi_{iJ}]. \tag{2.116}$$

2.8.2 *Piezoelectric sensor*

Consider the kth piezoelectric sensor patch attached to the beam. It is assumed that the piezoelectric sensor is placed on the top surface of the beam as shown in Figure 2.18. When the beam vibrates due to some external disturbance or non-zero initial conditions, strain is generated inside the beam. Consequently, an electric charge is generated inside the piezoelectric sensor due to the piezoelectric effect. The electric charge distribution $q_p(t)$, i.e. the charge per unit length, can be written as [91]

$$q_p(t) = \frac{k_{31}^2}{g_{31}}\, \epsilon_r\, W_p, \tag{2.117}$$

where k_{31} is the electromechanical coupling factor and g_{31} is the piezoelectric voltage constant in r direction. Using Hooke's Law for the beam deflection in r direction, the expression for the sensor strain is obtained as

$$\epsilon_r = -y_p \frac{\partial^2 y}{\partial r^2}, \tag{2.118}$$

where

$$y_p = -\frac{h + h_p}{2}$$

is the normal distance from the neutral-axis of the beam to the mid-plane of the sensor patch.

The overall electric charge generated can be obtained by integrating the expression $q_p(t)$ across the length of the sensor (2.117). Substituting the solution form into (2.112), the induced sensor voltage $V_{sk}(t)$, at the kth sensor is found to be

$$V_{sk}(t) = \Omega_k \sum_{i=1}^{\infty} \int_{r_{1k}}^{r_{2k}} \frac{d^2\phi_i(r)}{dr^2}\, dr\, q_i(t), \tag{2.119}$$

where

$$\Omega_k = \frac{W_{pk}\, k_{31k}^2}{C_k\, g_{31k}} \left(\frac{h + h_{pk}}{2} \right) \tag{2.120}$$

and C_k is the capacitance of the piezoelectric sensor. The integral given in the above equation can be shown to be equal to Ψ_{ij} in (2.114) when j is replaced by k. Thus, the contribution of each mode to the overall sensor voltage is proportional to Ψ_{ik}.

Taking the Laplace Transform of (2.113) and (2.119), the MIMO (Multiple-Input, Multiple-Output) transfer function from the applied actuator-voltages $V_a(s)$ to the sensor voltages $V_s(s) = [V_{s1}(s), \ldots, V_{sJ}(s)]'$ can be written as

$$G_{Vs}(s) = \sum_{i=1}^{\infty} \frac{\Upsilon_i P_i}{s^2 + 2\zeta_i \omega_i s + \omega_i^2}, \tag{2.121}$$

where $\Upsilon_i = [\Omega_1 \Psi_{i1}, \ldots, \Omega_J \Psi_{iJ}]'$. If all piezoelectric actuators and sensors are collocated, then $\Upsilon_i P_i$ is a positive semi-definite matrix. Similar transfer functions can be obtained for more complicated structures, such as plates with certain boundary conditions and tubular structures, by using numerical techniques such as finite element modeling.

2.9 Conclusions

In this chapter we studied the dynamics of a number of flexible structures with specific boundary conditions. We derived transfer functions that represent the behaviour of these systems, and demonstrated that the resulting models capture the spatially distributed nature of these systems. In the forthcoming chapters we will explain how such models can be utilized to design effective controllers capable of minimizing structural vibration in a global sense.

The majority of systems studied in this chapter were flexible structures such as beams and plates with regular shapes and well-defined boundary conditions, whose eigenvalue problems can be solved analytically. For more complex system, where such an analytic solution can not be found, we may use spatial system identification to obtain a model based on measured input-output data. This will be studied in Chapter 7.

Chapter 3

Spatial Norms and Model Reduction

3.1 Introduction

In the previous chapter we introduced a class of spatially distributed systems and explained how they could be modeled using the modal analysis technique. Throughout this book, we will be working with systems of this nature. In order to use these models in analysis and synthesis problems, we need to define suitable performance measures that take into account their spatially distributed nature. Traditional performance measures, such as $\mathcal{H}_2$ and $\mathcal{H}_\infty$ norms, only deal with point-wise models for such systems. In this chapter, we extend these performance measures to include the spatial characteristics of spatially distributed systems.

Another topic that is covered in this chapter is that of model reduction by balanced truncation. The problem of model reduction for dynamical systems has been studied extensively throughout the literature; see, for example, [26, 112, 30]. Here, we address the problem for spatially distributed linear time-invariant systems, following [73].

3.2 Spatial $\mathcal{H}_2$ norm

In the previous chapter we demonstrated that modeling of spatio-temporal systems of the form (2.1) using the modal analysis approach results in models of the form

$$G(s,r) = \sum_{i=1}^{\infty} \frac{\phi_i(r)}{s^2 + 2\zeta_i\omega_i s + \omega_i^2} H_i, \tag{3.1}$$

where r belongs to a known set, i.e., $r \in \mathcal{R}$. Notice that if the system has several inputs, then H_i, $i = 1, 2, \ldots$ will be row vectors with the same dimension as the number of inputs. Moreover, we argued that the orthogonality condition can often be written as

$$\int_{\mathcal{R}} \phi_i(r)\phi_j(r)dr = \Phi_i^2 \delta_{ij}. \tag{3.2}$$

This is true for a large number of systems. For instance, beam-like and plate-like structures with uniform mass distribution, acoustic enclosures with a uniform cross section and uniform strings satisfy this condition. It can be observed that (3.1) consists of an infinite number of orthogonal modes. Moreover, G describes the spatial as well as spectral behavior of the system.

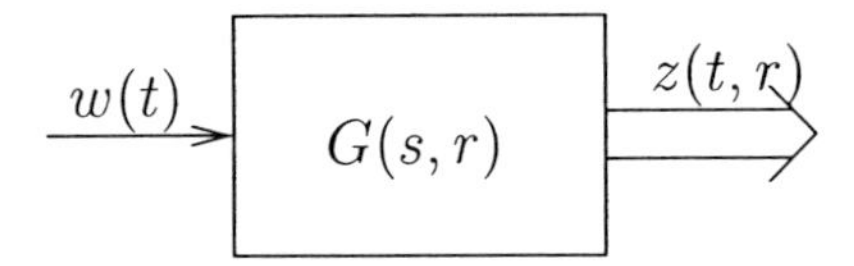

Fig. 3.1 A spatially distributed linear time-invariant system

In this chapter, we develop the requisite mathematical machinery that is needed for proving our main results in the remainder of the book. To this end, we consider a system of the form $G(s, r)$ that maps an input signal $w(t) \in \mathbf{R}^m$ to an output signal $z(t, r) \in \mathbf{R}^\ell \times \mathcal{R}$. The system has a finite number of inputs, however, its outputs are spatially distributed over a given set $\mathcal{R}$ (see Figure 3.1). We will need the following definitions.

Definition 3.1 Spatial $\mathcal{H}_2$ norm of a signal: Consider a signal $z(t, r) \in \mathbf{R}^\ell \times \mathcal{R}$. Then, the *spatial $\mathcal{H}_2$ norm* of z is defined as

$$\ll z \gg_2^2 = \int_0^\infty \int_{\mathcal{R}} z(t, r)' z(t, r) dr dt. \tag{3.3}$$

Spatial $\mathcal{H}_2$ norm of $z(t, r)$ can be interpreted as the total energy of the spatially distributed signal z.

Definition 3.2 Spatial $\mathcal{H}_2$ norm of a system: Consider a system $G(s, r)$ with $r \in \mathcal{R}$. The *spatial $\mathcal{H}_2$ norm* of this system is defined as

$$\ll G(s, r) \gg_2^2 = \frac{1}{2\pi} \int_{-\infty}^{\infty} \int_{\mathcal{R}} \text{tr}\{G(j\omega, r)^* G(j\omega, r)\} dr d\omega. \tag{3.4}$$

For a single input system, such as a beam with a single point force, $\ll G(s,r) \gg_2^2$ is a measure of the volume underneath the surface defined by $|G(j\omega,r)|^2$. Hence, this is a natural extension of the standard $\mathcal{H}_2$ norm of linear time-invariant systems to systems of the form (3.1). Similar interpretations can be made for transfer functions of plates, etc.

The question now is: given a system of the form (3.1), how can the spatial $\mathcal{H}_2$ norm as defined by (3.4) be computed? The following theorem contains the answer.

Theorem 3.1 *Consider $G(s,r)$ as defined in (3.1) and suppose that the mode shapes satisfy the orthogonality condition (3.2). Then,*

$$\ll G(s,r) \gg_2^2 = \sum_{i=1}^{\infty} \|\tilde{G}_i(s)\|_2^2, \tag{3.5}$$

where

$$\tilde{G}_i(s) = \frac{\Phi_i}{s^2 + 2\zeta_i\omega_i s + \omega_i^2} H_i. \tag{3.6}$$

Proof

$$\begin{aligned}
&\ll G(s,r) \gg_2^2 \\
&= \frac{1}{2\pi}\int_{-\infty}^{\infty}\int_{\mathcal{R}} \mathrm{tr}\left\{G(j\omega,r)^* G(j\omega,r)\right\} dr d\omega \\
&= \frac{1}{2\pi}\int_{-\infty}^{\infty}\int_{\mathcal{R}} \mathrm{tr}\Bigg\{\left(\sum_{k=1}^{\infty}\frac{\phi_k(r)}{-\omega^2 + j2\zeta_k\omega_k\omega + \omega_k^2} H_k'\right) \\
&\qquad\qquad \times \left(\sum_{l=1}^{\infty}\frac{\phi_l(r)}{-\omega^2 - j2\zeta_l\omega_l\omega + \omega_l^2} H_l\right)\Bigg\} dr d\omega \\
&= \frac{1}{2\pi}\int_{-\infty}^{\infty}\sum_{i=1}^{\infty}\left|\frac{\Phi_i}{-\omega^2 + j2\zeta_i\omega_i\omega + \omega_i^2}\right|^2 \mathrm{tr}\{H_i' H_i\} d\omega \\
&= \sum_{i=1}^{\infty}\|\tilde{G}_i\|_2^2.
\end{aligned}$$

□

The above theorem proves that the spatial $\mathcal{H}_2$ norm of a spatially distributed system of the form (3.1) with the orthogonality condition (3.2) is equivalent to the $\mathcal{H}_2$ norm of a finite-dimensional linear time-invariant system. The latter can be determined in a number of ways [112, 97].

Now, let us assume that the model (3.1) is truncated by keeping the first N modes only. In that case, $G(s,r)$ is approximated by

$$G_N(s,r) = \sum_{i=1}^{N} \frac{\phi_i(r)}{s^2 + 2\zeta_i\omega_i s + \omega_i^2} H_i. \tag{3.7}$$

If the result of Theorem 3.1 is applied to (3.7), we conclude that

$$\ll G_N \gg_2^2 = \sum_{i=1}^{N} \|\tilde{G}_i(s)\|_2^2.$$

An interpretation of this result is that the spatial $\mathcal{H}_2$ norm of $G_N(s,r)$ is equivalent to the $\mathcal{H}_2$ norm of the following MIMO system

$$\hat{G}(s) = \begin{bmatrix} \frac{\Phi_1}{s^2+2\zeta_1\omega_1 s+\omega_1^2} H_1 \\ \vdots \\ \frac{\Phi_N}{s^2+2\zeta_N\omega_N s+\omega_N^2} H_N \end{bmatrix}. \tag{3.8}$$

This can be easily verified. Also, notice that as the number of selected modes is increased, so does the number of outputs of $\hat{G}(s)$. If an infinity of modes are considered, then the transfer function $\hat{G}(s)$ will have an infinite number of outputs.

3.3 Spatial $\mathcal{H}_\infty$ norm

$\mathcal{H}_2$ norm is a widely used measure of performance for linear time-invariant systems. Another performance measure that is often used in analysis and synthesis problems is the $\mathcal{H}_\infty$ norm. In this section, we extend the notion of $\mathcal{H}_\infty$ norm to the spatially distributed linear time-invariant systems.

Definition 3.3 Spatial Induced norm of a system: Let $\mathcal{G}$ be the linear operator which maps the inputs of $G(s,r)$ to its outputs. The *spatial induced norm of* $\mathcal{G}$ is defined as

$$\ll \mathcal{G} \gg^2 = \sup_{0 \neq w \in \mathcal{L}_{2[0,\infty)}} \frac{\ll z \gg_2^2}{\|w\|_2^2}. \tag{3.9}$$

Definition 3.4 Spatial $\mathcal{H}_\infty$ norm of a system: Consider a spatially distributed linear time-invariant system $G(s,r)$. The *spatial* $\mathcal{H}_\infty$ *norm* of this system is defined as

$$\ll G \gg_\infty^2 = \sup_{\omega \in \mathbf{R}} \lambda_{\max} \left(\int_{\mathcal{R}} G(j\omega, r)^* G(j\omega, r) dr \right). \tag{3.10}$$

The following theorem proves that the spatial $\mathcal{H}_\infty$ norm of $G(s,r)$ is indeed equivalent to the spatial induced norm of $\mathcal{G}$.

Theorem 3.2 *Suppose a stable linear system has a transfer function matrix $G(s,r)$ and let $\mathcal{G}$ denote the linear map it induces from the $\mathcal{L}_2$ spaces of its inputs to its infinite-dimensional outputs. Its induced operator norm $\ll \mathcal{G} \gg$ satisfies*

$$\ll \mathcal{G} \gg = \ll G \gg_\infty .$$

Proof Let us write the linear relationship between the input and output as $z = \mathcal{G}w$. Applying Parseval's theorem, we have

$$\begin{aligned} &\int_0^\infty \int_{\mathcal{R}} z(t,r)' z(t,r) dr dt \\ &\quad = \frac{1}{2\pi} \int_{-\infty}^{\infty} \int_{\mathcal{R}} \hat{u}(j\omega)^* G(j\omega, r)^* G(j\omega, r) \hat{u}(j\omega) dr d\omega \\ &\quad \leq \frac{1}{2\pi} \int_{-\infty}^{\infty} \lambda_{\max} \left(\int_{\mathcal{R}} G(j\omega, r)^* G(j\omega, r) dr \right) \hat{u}(j\omega)^* \hat{u}(j\omega) d\omega. \end{aligned}$$

Since

$$\ll G \gg_\infty^2 = \sup_{\omega \in \mathbf{R}} \lambda_{\max} \left(\int_{\mathcal{R}} G(j\omega, r)^* G(j\omega, r) dr \right),$$

it follows that

$$\ll z \gg_2 \leq \ll G \gg_\infty \|w\|_2.$$

That is,

$$\ll \mathcal{G} \gg \leq \ll G \gg_\infty .$$

Hence, $\ll G \gg_\infty$ is an upper bound for $\ll \mathcal{G} \gg$. Now, in order to show that it is the least upper bound, we assume that $\ll G \gg_\infty > \gamma$. This implies that for some ω_0, we have

$$\lambda_{\max}\left(\int_{\mathcal{R}} G(j\omega_0, r)^* G(j\omega_0, r) dr\right) > \gamma.$$

Hence, by continuity there exists $\eta > 0$ such that

$$\lambda_{\max}\left(\int_{\mathcal{R}} G(j\omega, r)^* G(j\omega, r) dr\right) \geq \gamma$$

for all $\omega \in [\omega_0 - \eta, \omega_0 + \eta]$ and $[-\omega_0 - \eta, -\omega_0 + \eta]$. Now consider a $\hat{u}(j\omega)$ which is zero outside of these ranges of frequencies and coincides with an eigenvalue corresponding to the largest eigenvalue of $\int_{\mathcal{R}} G(j\omega, r)^* G(j\omega, r) dr$ in these ranges. For the corresponding output z we then have

$$\begin{aligned}
\ll z \gg_2^2 &= \int_0^\infty \int_{\mathcal{R}} z(t,r)' z(t,r) dr dt \\
&= \frac{1}{2\pi} \int_{-\omega_0-\eta}^{-\omega_0+\eta} \int_{\mathcal{R}} \hat{u}(j\omega)^* G(j\omega, r)^* G(j\omega, r) \hat{u}(j\omega) dr d\omega \\
&\quad + \frac{1}{2\pi} \int_{\omega_0-\eta}^{\omega_0+\eta} \int_{\mathcal{R}} \hat{u}(j\omega)^* G(j\omega, r)^* G(j\omega, r) \hat{u}(j\omega) dr d\omega \\
&\geq \frac{1}{2\pi} \int_{-\omega_0-\eta}^{-\omega_0+\eta} \gamma^2 \hat{u}(j\omega)^* \hat{u}(j\omega) d\omega \\
&\quad + \frac{1}{2\pi} \int_{\omega_0-\eta}^{\omega_0+\eta} \gamma^2 \hat{u}(j\omega)^* \hat{u}(j\omega) d\omega \\
&= \gamma^2 \|w\|_2^2.
\end{aligned}$$

Hence, $\ll \mathcal{G} \gg \geq \gamma$. This completes the proof of the theorem. □

3.4 Weighted spatial norms

In the previous sections we introduced the notions of spatial $\mathcal{H}_2$ and $\mathcal{H}_\infty$ norms for spatially distributed systems. These norms can be used as measures of performance in certain analysis and synthesis problems. In some cases, it can be beneficial to add a spatial weighting function to emphasize certain regions within the set $\mathcal{R}$. In this section, we extend those definitions to allow for spatially distributed weighting functions.

Definition 3.5 Weighted spatial $\mathcal{H}_2$ norm of a signal: Consider a signal $z(t,r) \in \mathbf{R}^\ell \times \mathcal{R}$. Then, the *weighted spatial $\mathcal{H}_2$ norm* of z is defined as

$$\ll z \gg_{2,Q}^{2} = \int_{0}^{\infty} \int_{\mathcal{R}} z(t,r)' Q(r) z(t,r) dr dt. \tag{3.11}$$

Notice that if $Q(r)$ is chosen to be a Dirac delta function at a point $r_a \in \mathcal{R}$, i.e., $Q(r) = \delta(r - r_a)$ then (3.11) reduces to:

$$\ll z \gg_{2,Q}^{2} = \int_{0}^{\infty} z(t,r_a)' z(t,r_a) dt$$

which is the $\mathcal{H}_2$ norm of the signal z at the point r_a. A similar discussion applies to the following definitions.

Definition 3.6 Weighted spatial $\mathcal{H}_2$ norm of a system: Consider a system $G(s,r)$ with $r \in \mathcal{R}$. The *weighted spatial $\mathcal{H}_2$ norm* of this system is defined as

$$\ll G(s,r) \gg_{2,Q}^{2} = \frac{1}{2\pi} \int_{-\infty}^{\infty} \int_{\mathcal{R}} \operatorname{tr}\{G(j\omega,r)^* Q(r) G(j\omega,r)\} dr d\omega. \tag{3.12}$$

Definition 3.7 Weighted spatial induced norm of a system: Let $\mathcal{G}$ be the linear operator which maps the inputs of $G(s,r)$ to its outputs. The *weighted spatial induced norm of $\mathcal{G}$* is defined as

$$\ll \mathcal{G} \gg^{2} = \sup_{0 \neq w \in \mathcal{L}_{2[0,\infty)}} \frac{\ll z \gg_{2,Q}^{2}}{\|w\|_2^2}. \tag{3.13}$$

Definition 3.8 Weighted spatial $\mathcal{H}_\infty$ norm of a spatially distributed linear time-invariant system: Consider a system $G(s,r)$. The *weighted spatial $\mathcal{H}_\infty$ norm* of this system is defined as

$$\ll G \gg_{\infty,Q}^{2} = \sup_{\omega \in \mathbf{R}} \lambda_{\max} \left(\int_{\mathcal{R}} G(j\omega,r)^* Q(r) G(j\omega,r) dr \right). \tag{3.14}$$

It is straightforward to show that Theorem 3.2 holds for weighted spatial norms. That is,

$$\ll \mathcal{G} \gg_{Q} = \ll G \gg_{\infty,Q}.$$

3.5 State-space forms

Transfer functions of the systems that are of concern here are generally described by (3.1). These models are often approximated by truncating high frequency modes. This results in models of the form (3.7). A state space realization of such a system can be shown to be

$$\begin{aligned} \dot{x}(t) &= Ax(t) + Bu(t) \\ y(t,r) &= C(r)x(t), \end{aligned} \tag{3.15}$$

where

$$A = \begin{bmatrix} 0 & 1 & 0 & 0 \\ -\omega_1^2 & -2\zeta_1\omega_1 & 0 & 0 \\ & & \ddots & \\ 0 & 0 & 0 & 1 \\ 0 & 0 & -\omega_N^2 & -2\zeta_N\omega_N \end{bmatrix}; \quad B = \begin{bmatrix} 0 \\ H_1 \\ \vdots \\ 0 \\ H_N \end{bmatrix}$$

$$C(r) = \begin{bmatrix} \phi_1(r) \ 0 \ \dots \ \phi_N(r) \ 0 \end{bmatrix}$$

and the state vector $x(t)$ is defined in terms of modal coordinates (see Section 2.2), i.e.

$$x(t) = \begin{bmatrix} q_1(t) \\ \dot{q}_1(t) \\ \vdots \\ q_N(t) \\ \dot{q}_N(t) \end{bmatrix}.$$

In the next theorem, we show that the spatial $\mathcal{H}_2$ norm of a system of the form (3.15) is equivalent to the $\mathcal{H}_2$ norm of a finite-dimensional system. The importance of this theorem is that it allows the standard computational tools to be used in calculating the spatial $\mathcal{H}_2$ norm of spatially distributed systems.

Theorem 3.3 *Consider a stable linear system with a transfer function matrix $G(s,r)$ that can be described in state-space form by (3.15). Then*

$$\ll G(s,r) \gg_2 = \|\tilde{G}(s)\|_2,$$

where

$$\tilde{G}(s) = \Gamma(sI - A)^{-1}B$$

and

$$\Gamma'\Gamma = \int_{\mathcal{R}} C(r)'C(r)dr. \tag{3.16}$$

Proof The proof is straightforward. □

In most of the problems that we will encounter in this book, the integral (3.16) can be easily evaluated using the orthogonality conditions. Otherwise, it can be determined numerically.

A similar result can be proved for the spatial $\mathcal{H}_\infty$ norm of a system of the form (3.15). However, to make the discussion more general, we can allow for a spatially-variant feed-through term as follows

$$\begin{aligned} \dot{x}(t) &= Ax(t) + Bu(t) \\ y(t,r) &= C(r)x(t) + D(r)u(t). \end{aligned} \tag{3.17}$$

In the following chapter, we will thoroughly explain why it may be crucial to include a feed-through term in the truncated models of structures. To this end, however, we point out that by truncating higher frequency modes of structures, we introduce an error at DC. In other words, the truncated modes of a structure do contribute to the low-frequency dynamics of the system. In the next chapter, we will show that this effect can be captured by adding a spatially distributed term to the truncated model of the structure.

Theorem 3.4 *Consider a stable linear system with a transfer function matrix $G(s,r)$ that can be described in state-space form by (3.17). Then*

$$\ll G(s,r) \gg_\infty = \|\tilde{G}(s)\|_\infty,$$

where

$$\tilde{G}(s) = \Pi(sI - A)^{-1}B + \Phi$$

and

$$\begin{bmatrix}\Pi & \Phi\end{bmatrix} = \Gamma,$$

where

$$\Gamma'\Gamma = \int_{\mathcal{R}} \begin{bmatrix} C(r)' \\ D(r)' \end{bmatrix} \begin{bmatrix} C(r) \; D(r) \end{bmatrix} dr. \tag{3.18}$$

Proof The proof follows from Definition 3.3, Theorem 3.2 and routine algebra. □

3.6 The balanced realization and model reduction by truncation

We have seen that models of distributed systems consist of an infinite number of modes. Furthermore, we have seen that these models are often simplified by truncating high-frequency modes. This is a poor way of reducing the order of a system. A better alternative is model reduction using balanced realization [30]. In this section we develop a model reduction methodology based on the balanced realization of a spatially distributed linear time-invariant system.

We consider a system whose state-space model is represented by (3.17). We will show that a reduced order model of the system can be obtained by balanced truncation of (3.17). Furthermore, we will show that such a truncation results in a bound on the spatial $\mathcal{H}_\infty$ norm of the error system. But first, we define what we mean by a balanced realization of (3.17).

Definition 3.9 A realization $(A, B, C(r))$ is balanced if A is a stability matrix and

$$A\Sigma + \Sigma A' + BB' = 0 \tag{3.19}$$

$$A'\Sigma + \Sigma A + R'R = 0, \tag{3.20}$$

where

$$\Sigma = diag(\sigma_1 I_{s_1}, \ldots, \sigma_N I_{s_N})$$

such that $\sigma_1 > \sigma_2 > \ldots > \sigma_N > 0$ and

$$R'R = \int_{\mathcal{R}} C(r)'Q(r)C(r)dr.$$

Notice that a spatial weighting function is inserted in the above equation. Hence, we are dealing with the weighted spatial norms here.

Now, let us assume that $(A, B, C(r))$ is a balanced realization as defined above and $\Sigma = diag(\Sigma_1, \Sigma_2)$, where

$$\Sigma_1 = diag(\sigma_1 I_{s_1}, \ldots, \sigma_m I_{s_m})$$
$$\Sigma_2 = diag(\sigma_{m+1} I_{s_{m+1}}, \ldots, \sigma_N I_{s_N}). \tag{3.21}$$

Here we are assuming that the system is of order N and it is to be approximated by a spatially distributed linear time-invariant model of order $m < N$.

Let us partition A, B and $C(r)$ matrices conformably with Σ as follows

$$A = \begin{bmatrix} A_{11} & A_{12} \\ A_{21} & A_{22} \end{bmatrix}; \quad B = \begin{bmatrix} B_1 \\ B_2 \end{bmatrix};$$
$$C(r) = \begin{bmatrix} C_1(r) & C_2(r) \end{bmatrix}. \tag{3.22}$$

Then, $(A_{11}, B_1, C_1(r), D(r))$ is said to be a balanced truncation of the system $(A, B, C(r), D(r))$.

Remark 3.1 *If the system (3.17) is controllable and observable, in the sense that the Lyapunov equations*

$$AP + PA' + BB' = 0$$

and

$$A'Q + QA + R'R = 0$$

have positive-definite, and symmetric solutions, then a similarity matrix T can be found such that $TPT' = T'QT = \Sigma$. Reference [30] suggests a procedure to construct such a similarity matrix.

The following lemma is important in the proof of the main result of this section.

Lemma 3.1 *Suppose that $(A, B, C(r))$ is a balanced realization as described above and that $(A_{11}, B_1, C_1(r), D(r))$ is a balanced truncation of $(A, B, C(r), D(r))$. Then $(A_{11}, B_1, C_1(r))$ is a balanced realization.*

Proof Let us assume that

$$R'R = \int_0^l \begin{bmatrix} C_1(r)' \\ C_2(r)' \end{bmatrix} Q(r) \begin{bmatrix} C_1(r) & C_2(r) \end{bmatrix} dr \tag{3.23}$$

and

$$[J \;\; K] = R,$$

where J is a $N \times m$ matrix and K is a $N \times (N-m)$ matrix.

Equations (3.19) and (3.20) imply that

$$A_{11}\Sigma_1 + \Sigma_1 A'_{11} + B_1 B'_1 = 0, \tag{3.24}$$

$$A'_{11}\Sigma_1 + \Sigma_1 A_{11} + J'J = 0, \tag{3.25}$$

$$A_{21}\Sigma_1 + \Sigma_2 A'_{12} + B_2 B'_1 = 0, \tag{3.26}$$

$$\Sigma_2 A_{21} + A'_{12}\Sigma_1 + K'J = 0, \tag{3.27}$$

$$A_{22}\Sigma_2 + \Sigma_2 A'_{22} + B_2 B'_2 = 0 \text{ and} \tag{3.28}$$

$$\Sigma_2 A_{22} + A'_{22}\Sigma_2 + K'K = 0. \tag{3.29}$$

If we can show that A_{11} is a stability matrix, then it immediately follows that $(A_{11}, B_1, C_1(r))$ is a balanced realization since $\Sigma_1 > 0$.

To prove that $Re(\lambda_i(A_{11})) < 0$, we only have to show that A_{11} cannot have an imaginary axis eigenvalue. For $\Sigma_1 > 0$ in (3.24) implies that $Re(\lambda_i(A_{11})) \leq 0$. The proof is based on contradiction.

Let us suppose that there exists a ω such that $(j\omega I - A_{11})$ is singular. Let $V \in Ker(j\omega I - A_{11})$. That is, $(j\omega I - A_{11})V = 0$. Now, if we multiply (3.25) on the left by V^* and on the right by V, we obtain $JV = 0$, and if we multiply it on the right by V, we obtain $(j\omega I + A'_{11})\Sigma_1 V = 0$. In the same manner, if we multiply (3.24) on the left by $V^*\Sigma_1$ and on the right by $\Sigma_1 V$, we obtain $B'_1\Sigma_1 V = 0$, and if we multiply it on the right by $\Sigma_1 V$, we obtain $(j\omega I - A_{11})\Sigma_1^2 V = 0$.

Therefore, $\Sigma_1^2 V \in Ker(j\omega I - A_{11})$, and hence, a matrix Σ_1^2 can be found such that $\Sigma_1^2 V = V\bar{\Sigma}_1^2$ and $\rho(\bar{\Sigma}_1^2) \subset \rho(\Sigma_1^2)$, where $\rho(M)$ is the spectrum of the matrix M.

Now, multiply (3.26) on the right by $\Sigma_1 V$ and multiply (3.27) on the left by Σ_2 and on the right by V to obtain

$$A_{21}\Sigma_1^2 V + \Sigma_2 A'_{12}\Sigma_1 V = 0 \text{ and}$$
$$\Sigma_2^2 A_{21} V + \Sigma_2 A'_{12}\Sigma_1 V = 0,$$

and then subtract them to obtain

$$\Sigma_2^2 A_{21} V = A_{21}\Sigma_1^2 V = A_{21} V \bar{\Sigma}_1^2.$$

This can be written as

$$\begin{bmatrix} \bar{\Sigma}_1^2 & 0 \\ 0 & \Sigma_2^2 \end{bmatrix} \begin{bmatrix} I \\ A_{21}V \end{bmatrix} = \begin{bmatrix} I \\ A_{21}V \end{bmatrix} \bar{\Sigma}_1^2.$$

Since $\bar{\Sigma}_1^2$ and Σ_2^2 have no eigenvalues in common, $[I \;\; (A'_{21}V)']'$ is a basis for the eigenspace of

$$\begin{bmatrix} \bar{\Sigma}_1^2 & 0 \\ 0 & \Sigma_2^2 \end{bmatrix}$$

corresponding to the eigenvalues of $\bar{\Sigma}_1^2$. Therefore,

$$\begin{bmatrix} I \\ A_{21}V \end{bmatrix} = \begin{bmatrix} I \\ 0 \end{bmatrix}.$$

This implies that $A_{21}V = 0$, which along with our earlier assumption that $(j\omega I - A)V = 0$ imply that

$$(j\omega I - A)\begin{bmatrix} I \\ 0 \end{bmatrix} = 0.$$

This is in contradiction with the assumption that A is a stability matrix. This proves that $Re(\lambda_i(A_{11})) < 0$ and therefore, the theorem. □

The next theorem is the main result of this section. It shows that a balanced truncation of system (3.17) results in an upper bound on the spatial $\mathcal{H}_\infty$ norm of the error system.

Theorem 3.5 *Suppose $G(s,r) = C(r)(sI - A)^{-1}B + D(r)$ is a balanced realization with Gramian $\Sigma = diag(\Sigma_1, \Sigma_2)$ as in (3.21) and assume that the $(A, B, C(r))$ is partitioned as in (3.22). Then the truncated system $G_m(s,r) = C_1(r)(sI - A_{11})^{-1}B_1 + D(r)$ is balanced (and hence asymptotically stable). Furthermore,*

$$\ll G(s,r) - G_m(s,r) \gg_{\infty,Q} \leq 2(\sigma_{m+1} + \sigma_{m+2} + \ldots + \sigma_N).$$

Moreover, the bound is achieved if $m = N - 1$, i.e.,

$$\ll G(s,r) - G_m(s,r) \gg_{\infty,Q} = 2\sigma_N.$$

Proof Let us consider the finite-dimensional systems

$$\tilde{G}(s) = R(sI - A)^{-1}B + \tilde{D}$$

and

$$\tilde{G}_m(s) = J(sI - A_{11})^{-1}B_1 + \tilde{D},$$

where $R = [J \;\; K]$ as in (3.23) and A and B matrices are partitioned as in (3.22) and $\tilde{D}$ is a matrix with appropriate dimensions.

Now, consider the error system

$$E(s,r) = G(s,r) - G_m(s,r).$$

It can be verified that $E(s,r)$ has a state-space description

$$E(s,r) = C_e(r)(sI - A_e)^{-1}B_e,$$

where

$$A_e = \begin{bmatrix} A_{11} & 0 & 0 \\ 0 & A_{11} & A_{12} \\ 0 & A_{21} & A_{22} \end{bmatrix}; \quad B_e = \begin{bmatrix} B_1 \\ B_1 \\ B_2 \end{bmatrix}; \text{ and}$$

$$C_e(r) = \begin{bmatrix} -C_1(r) & C_1(r) & C_2(r) \end{bmatrix}. \tag{3.30}$$

It is straightforward, as well, to show that the finite-dimensional error system

$$\tilde{E}(s) = \tilde{G}(s) - \tilde{G}_m(s)$$

has a state-space description

$$\tilde{E}(s) = \tilde{C}_e(sI - A_e)^{-1}B_e,$$

where

$$\tilde{C}_e = \begin{bmatrix} -J & J & K \end{bmatrix}.$$

Theorem 3.2 implies that

$$\ll E(s,r) \gg_{\infty,Q} = ||\tilde{E}(s)||_\infty.$$

Moreover, it is obvious that $\tilde{G}(s)$ and $\tilde{G}_m(s)$ are finite-dimensional balanced realizations with the same Gramians as those of $G(s,r)$ and $G_m(s,r)$. Therefore, Theorem 7.3 of [112] implies that

$$||\tilde{G}(s) - \tilde{G}_m(s)||_\infty \leq 2(\sigma_{m+1} + \sigma_{m+2} + \ldots + \sigma_N)$$

and

$$||\tilde{G}(s) - \tilde{G}_{N-1}(s)||_\infty = 2\sigma_N.$$

This completes the proof of the theorem since we have shown that

$$\ll G(s,r) - G_m(s,r) \gg_{\infty,Q} = ||\tilde{G}(s) - \tilde{G}_m(s)||_\infty.$$

□

An implication of the above discussion is that in order to determine the spatial $\mathcal{H}_\infty$ norm of a system of the form (3.17), one needs to evaluate integrals of the the form

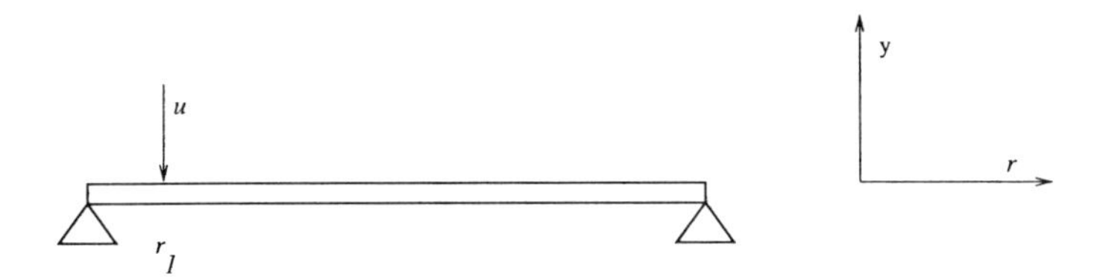

Fig. 3.2 A simply-supported beam subject to a point force disturbance

$$\int_{\mathcal{R}} \begin{bmatrix} C(r)' \\ D(r)' \end{bmatrix} Q(r) \begin{bmatrix} C(r) & D(r) \end{bmatrix} dr.$$

At first, this might seem to be a difficult task. However, thanks to the orthogonality conditions of the mode shapes, such an integral can be determined very easily. Consider the simply-supported beam of Figure 3.2. We know that the mode shapes satisfy the orthogonality condition (2.55). Therefore, assuming that $Q(r) = 1$ for $r \in [0, L]$,

$$\begin{aligned}
\Gamma'\Gamma &= \int_0^L C(r)'C(r)dr \\
&= \int_0^L \begin{bmatrix} \phi_1(r_1)\phi_1(r) \\ 0 \\ \vdots \\ \phi_N(r_1)\phi_N(r) \\ 0 \end{bmatrix} [\phi_1(r_1)\phi_1(r) \;\; 0 \;\; \dots \;\; \phi_N(r_1)\phi_N(r) \;\; 0]\, dr \\
&= \frac{1}{\rho A} \begin{bmatrix} \phi_1(r_1)^2 & 0 & & 0 & 0 \\ 0 & 0 & & 0 & 0 \\ & & \ddots & & \\ 0 & 0 & & \phi_N(r_1)^2 & 0 \\ 0 & 0 & & 0 & 0 \end{bmatrix}.
\end{aligned}$$

Therefore, corresponding to system (3.15), there exists the finite-dimensional system,

$$\begin{aligned}
\dot{x}(t) &= Ax(t) + Bu(t) \\
y(t) &= \Gamma x(t),
\end{aligned}$$

where Γ is the following $N \times 2N$ matrix:

$$\Gamma = \frac{1}{\sqrt{\rho A}} \begin{bmatrix} \phi_1(r_1) & 0 & & & & & \\ 0 & 0 & \phi_2(r_1) & 0 & & & \\ & & 0 & 0 & & & \\ & & & & \ddots & & \\ & & & & & \phi_2(r_1) & 0 \\ & & & & & 0 & 0 \end{bmatrix}.$$

It was proved, in the previous section, that this system has a $\mathcal{H}_\infty$ norm equivalent to the spatial $\mathcal{H}_\infty$ norm of the system (3.15). At this stage, we point out that in the majority of problems arising in flexible structures, we only need to assume that $Q(r) = 1$. However, if $Q(r)$ is chosen to be an arbitrary function of position, then calculating the matrix Γ could become a difficult analytic task. Nevertheless, such an integration can be carried out numerically.

3.7 Illustrative example

In this section, we apply our model reduction procedure to a simply-supported beam as shown in Figure 3.2. The beam parameters are as follows.

$$\begin{aligned} L &= \text{Beam length} = 0.38\ m, \\ r_1 &= 0.038\ m, \\ \rho A &= 0.6265\ kg/m \ \text{ and} \\ EI &= 5.329\ Nm^2. \end{aligned}$$

This is the beam described in [2]. We also assume a damping ratio of $\zeta_i = 0.01$ for each mode. In our simulations we consider the first six modes of the beam. The 3D frequency response of the beam can be observed in Figure 3.3.

This corresponds to a 12th order system. We apply our model reduction to this system to find a reduced order model of the 4th order. The 3D frequency response of the reduced order model is plotted in Figure 3.4. Theorem 3.5 guarantees an upper bound of 1.1788×10^{-4}. The actual $\mathcal{H}_\infty$ norm of the error system, however, is 7.342×10^{-5}. This is expected as Theorem 3.5 only gives an upper bound on the $\mathcal{H}_\infty$ norm of the error system. Figure 3.5 shows the 3D frequency plot of the error system.

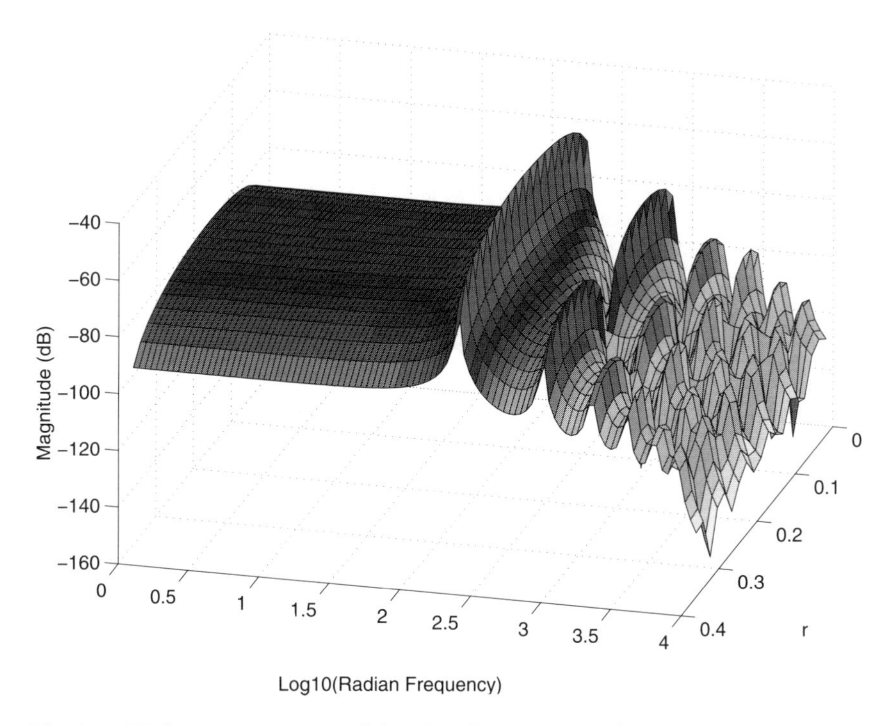

Fig. 3.3 3D frequency response of the pinned-pinned beam from the point force input (N) to the displacement (m)

Instead of using the above procedure to reduce the dimension of the 12th order beam model to a 4th order model, we could simply consider the first two modes of the beam. 3D frequency response of this model is plotted in Figure 3.6. Such a model would result in an error of 7.347×10^{-5} which is higher than the one obtained using the balanced model reduction procedure (Figure 3.7). Therefore, a balanced model reduction results in a reduced model of the beam which is closer to the higher order model in a spatial $\mathcal{H}_\infty$ norm sense.

It should be clear by now that transfer functions of flexible structures, such as a beam, consist of complex conjugate poles. Therefore, state-space models of such systems are of even order. An advantage of using balanced model reduction techniques is that the system could be approximated by a lower order system of odd dimension.

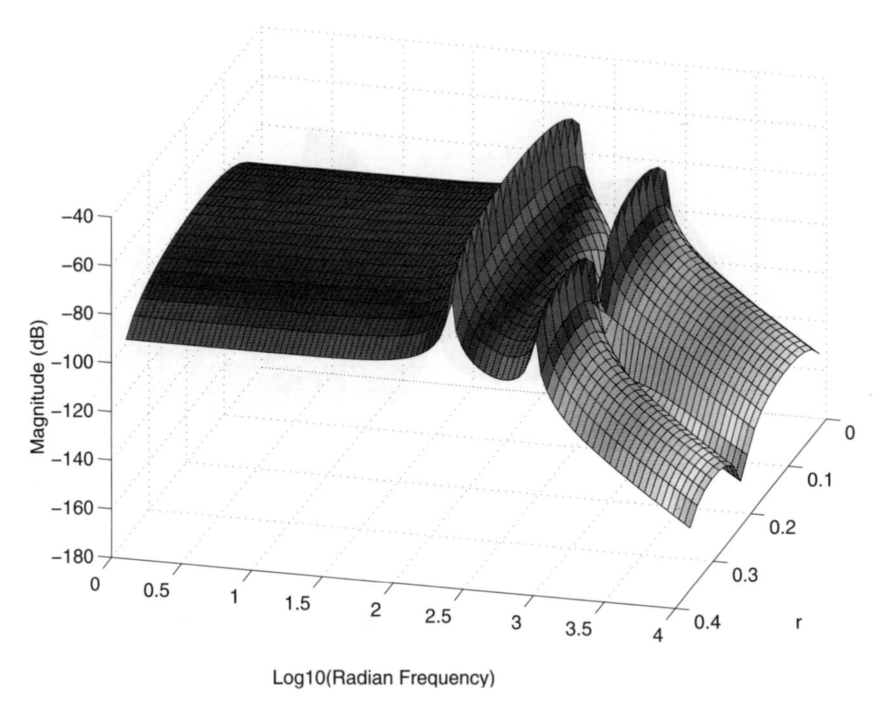

Fig. 3.4 3D frequency response of the reduced order model of a pinned-pinned beam from the point force (N) to the displacement (m)

3.8 Conclusions

Notions of spatial $\mathcal{H}_2$ and spatial $\mathcal{H}_\infty$ norms were introduced in this chapter. These measures are central to the theory that will be developed in the remainder of the book. Numerical calculations of these measures for a spatially distributed system were explained, and the techniques of balanced realization and model reduction by truncation were extended to the spatially distributed systems.

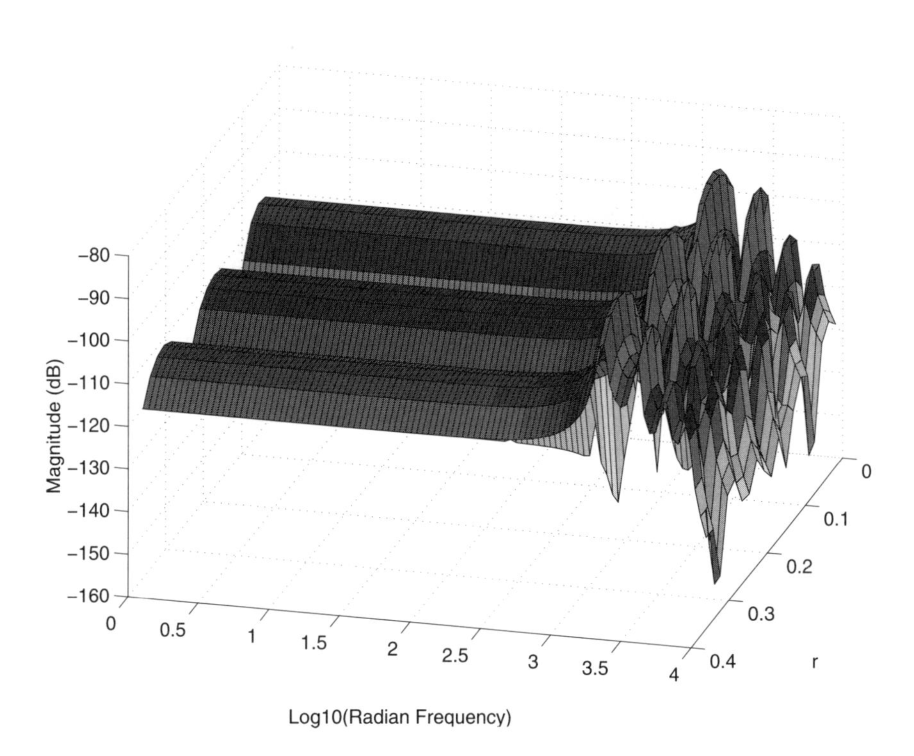

Fig. 3.5 3D frequency response of the error system

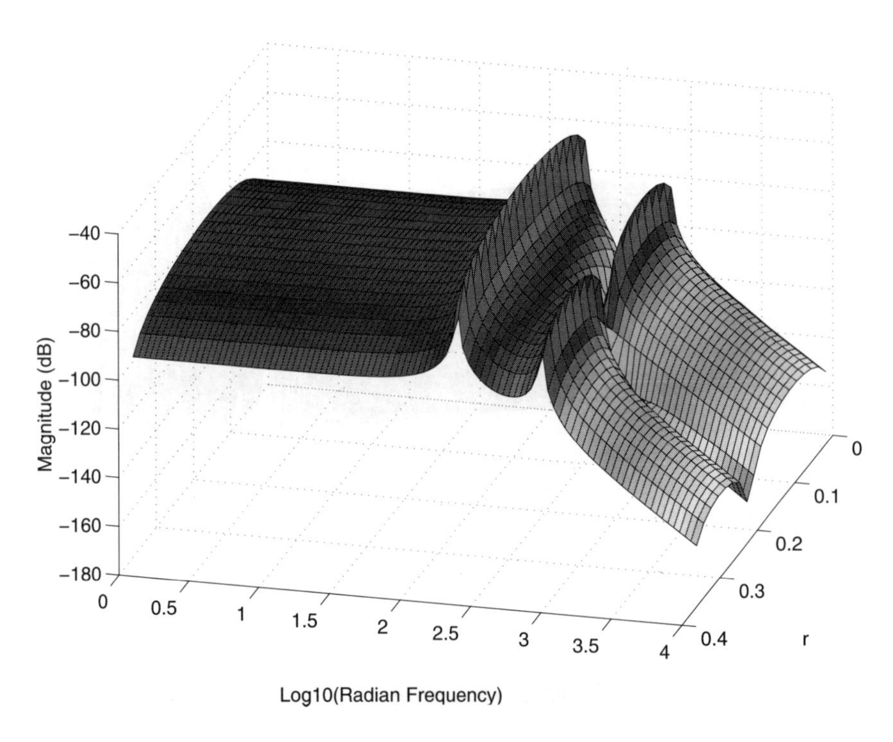

Fig. 3.6 3D frequency response of the beam with (only the first two modes included)

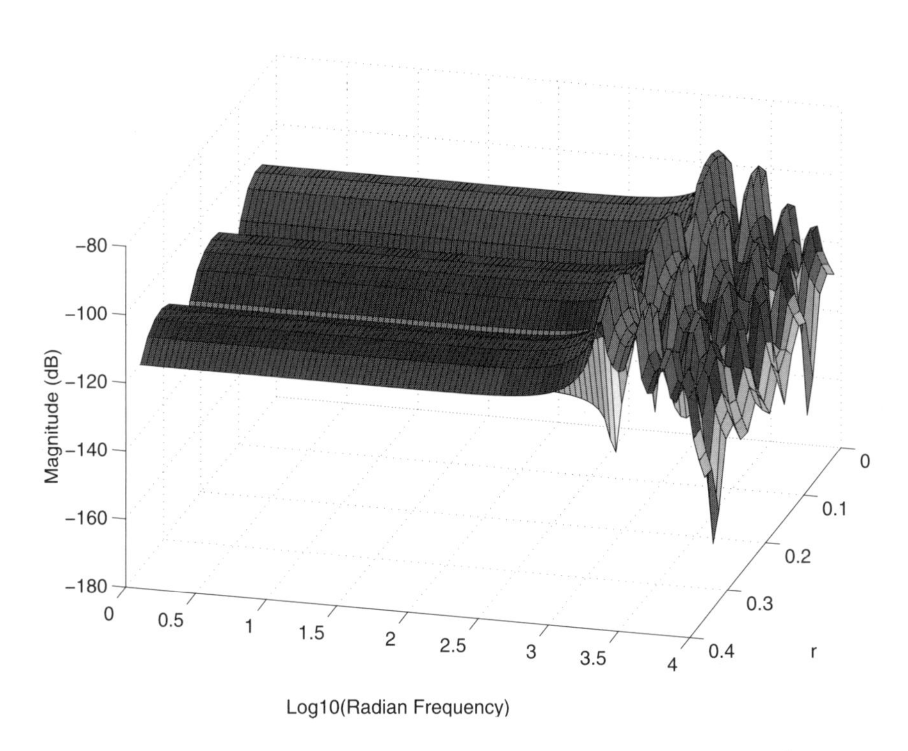

Fig. 3.7 3D frequency response of the error system

Chapter 4

Model Correction

4.1 Introduction

This chapter is concerned with the problem of model correction for the class of spatially distributed systems introduced in Chapter 2. Dynamics of flexible structures and some acoustic systems consist of an infinite number of modes. For control design purposes these models are often approximated by finite-dimensional models via truncation. Direct truncation of higher order modes results in the perturbation of zero dynamics and hence generates errors in the spatial frequency response of the system. If such a model is used to design a controller, and subsequently, the controller is implemented on the system, one may observe unexpected results. This is due to the fact that closed loop performance of the system is largely dictated by open loop zeros.

The model correction method that is proposed in this chapter allows for a spatially distributed DC term to capture the effect of truncated modes in an optimal way.

The first part of this chapter is concerned with the problem of model correction for spatially distributed systems. In the second part of the chapter we will concentrate on point-wise models of structures. In a typical control design problem, the spatial characteristics of the system are used to construct spatial performance indexes, while the point-wise model is needed for feedback.

4.2 Effect of truncation

Consider the single-input spatio-temporal system

$$G(s,r) = \sum_{i=1}^{\infty} \frac{\phi_i(r)F_i}{s^2 + \omega_i^2} \tag{4.1}$$

and, associated with that, the orthogonality condition

$$\int_{\mathcal{R}} \phi_i(r)\phi_j(r)dr = \Phi_i^2 \delta_{ij}. \tag{4.2}$$

The control designer is often interested only in designing a controller for a particular bandwidth. Therefore, an approximate model of the system is needed that best represents the dynamics of the system in the prescribed frequency range. In other words, a lower order dynamical model is needed. A first choice in this case might be to simply ignore the modes which correspond to the frequencies that lie outside the bandwidth of interest. For instance, if ω_N is equivalent or larger than the highest frequency of interest, one may choose to approximate $G(s,r)$ by

$$G_N(s,r) = \sum_{i=1}^{N} \frac{F_i\phi_i(r)}{s^2 + \omega_i^2}. \tag{4.3}$$

This appears to be the mainstream approach in simplifying the dynamics of this class of systems [11]. A drawback of this approximation is that the truncated higher order modes may contribute to the low-frequency dynamics, mainly in the form of distorting zero locations. Furthermore, these removed modes can significantly distort the spatial characteristics of the low-order model. Therefore, an approximate low-order model is needed that best captures the effect of truncated modes on the spectral (hence temporal) and spatial dynamics of the system.

To see how this truncation may perturb low-frequency dynamics, we look at the error system generated by approximating the full-order system (4.1) with the truncated model (4.3). That is,

$$\begin{aligned} E(s,r) &= G(s,r) - G_N(s,r) \\ &= \sum_{i=N+1}^{\infty} \frac{F_i\phi_i(r)}{s^2 + \omega_i^2}. \end{aligned}$$

At DC this amounts to a spatial error of

$$K(r) = \sum_{i=N+1}^{\infty} \frac{F_i \phi_i(r)}{\omega_i^2}. \tag{4.4}$$

This could introduce a significant amount of error at low frequencies, particularly if there exists a point r_a at which $F_i\phi_i(r_a) > 0$ for all i. This may happen if a point force is acting on the structure at r_a. In such a case the DC error can be significantly large at r_a. Therefore, it is sensible to expect that this error can be reduced if $K(r)$ is added to the truncated model (4.3). Indeed, the resulting corrected model

$$\hat{G}(s,r) = \sum_{i=1}^{N} \frac{F_i \phi(r)}{s^2 + \omega_i^2} + \sum_{i=N+1}^{\infty} \frac{F_i \phi_i(r)}{\omega_i^2} \tag{4.5}$$

will generate zero error at $\omega = 0$. However, the error will exist elsewhere within the bandwidth of interest. In the aeroelasticity literature this technique is referred to as the mode acceleration method (see page 350 of [5]) and is often applied to point-wise models of structures (see [11]).

It should be pointed out that $K(r)$ is often evaluated using a finite, but possibly large, number of terms, i.e.,

$$K(r) = \sum_{i=N+1}^{M} \frac{F_i \phi_i(r)}{\omega_i^2}.$$

This is quite reasonable since mode shapes are bounded functions of r and $\omega_i \to \infty$ as $i \to \infty$.

4.3 Model correction using the spatial $\mathcal{H}_2$ norm

In this section we follow the model correction technique that was proposed in the previous section. However, we attempt to find the correction term such that the corrected model approximates the real system in an optimal sense. Here, we approximate $G(s,r)$ by

$$\hat{G}(s,r) = G_N(s,r) + K(r),$$

where

$$K(r) = \sum_{i=N+1}^{\infty} k_i \phi_i(r). \tag{4.6}$$

Our intention is to find an optimal value for K such that the effect of truncated higher order modes on the low-frequency dynamical model of the system is minimized and the spatial characteristics of the model is best preserved within the bandwidth of interest.

Our approach to enhancing the approximation made by truncating the higher order modes of (4.1) is to add a spatially distributed feed-through term to the first N modes of (4.1). That is, to approximate (4.1) by

$$\hat{G}(s,r) = \sum_{i=1}^{N} \frac{F_i \phi_i(r)}{s^2 + \omega_i^2} + \sum_{i=N+1}^{\infty} k_i \phi_i(r). \tag{4.7}$$

This choice of the zero-frequency term is inspired by (4.5). However, unlike (4.5) we find the parameters k_i, $i = N+1, N+2, \ldots$ such that the following spatial cost function is minimized.

$$J = \ll (G(s,r) - \hat{G}(s,r))W(s,r) \gg_2^2 \tag{4.8}$$

Here, G and $\hat{G}$ are defined as in (4.1) and (4.7) and $W(s,r)$ is an ideal low-pass weighting function distributed spatially over $\mathcal{R}$ with its cut-off frequency ω_c chosen to lie within the interval $\omega_c \in (\omega_N, \omega_{N+1})$. That is,

$$|W(j\omega, r)| = \begin{cases} 1 & -\omega_c \leq \omega \leq \omega_c, \ r \in \mathcal{R} \\ 0 & \text{elsewhere.} \end{cases} \tag{4.9}$$

The reason for this choice of W will become clear soon. To this end, it should be clear that k_is chosen to minimize (4.8) will reduce the effect of out of bandwidth dynamics of $G(s,r)$ on $\hat{G}(s,r)$ in a spatial $\mathcal{H}_2$ optimal sense. Therefore, guaranteeing that the reduced order model $\hat{G}$ will best represent the frequency response of G while preserving its spatial characteristics.

Theorem 4.1 *Consider the system defined by (4.1), its approximation (4.7) and the spatial cost function (4.8). The parameters k_i, $i = N + 1, N+2, \ldots$ that minimize (4.8) are given by:*

$$k_i^{opt} = \frac{F_i}{2\omega_c \omega_i} \ln\left(\frac{\omega_i + \omega_c}{\omega_i - \omega_c}\right) \quad \textit{and} \quad i = N+1, N+2, \ldots . \tag{4.10}$$

Proof

It can be verified that (4.8) is equivalent to

$$J = \ll \left(\sum_{i=N+1}^{\infty} \frac{F_i \phi_i(r)}{s^2 + \omega_i^2} - \sum_{i=N+1}^{\infty} k_i \phi_i(r) \right) W(s,r) \gg_2^2 . \tag{4.11}$$

The fact that W is chosen to be an ideal spatial low-pass filter with its cut-off frequency lower than the first out-of-bandwidth pole of G, i.e. ω_{N+1}, guarantees that (4.11) will remain finite. The cost function (4.11) is then equivalent to:

$$\begin{aligned}
J &= \frac{1}{2\pi} \int_{-\omega_c}^{\omega_c} \int_{\mathcal{R}} \left| \sum_{i=N+1}^{\infty} \frac{F_i \phi_i(r)}{\omega_i^2 - \omega^2} \right|^2 dr d\omega \\
&\quad -2 \left(\frac{1}{2\pi} \int_{-\omega_c}^{\omega_c} \int_{\mathcal{R}} \left(\sum_{i=N+1}^{\infty} \frac{F_i \phi_i(r)}{\omega_i^2 - \omega^2} \right) \left(\sum_{i=N+1}^{\infty} k_i \phi_i(r) \right) \right) dr d\omega \\
&\quad + \frac{1}{2\pi} \int_{-\omega_c}^{\omega_c} \int_{\mathcal{R}} \left(\sum_{i=N+1}^{\infty} k_i \phi_i(r) \right)^2 dr d\omega \\
&= \frac{1}{2\pi} \int_{-\omega_c}^{\omega_c} \int_{\mathcal{R}} \left| \sum_{i=N+1}^{\infty} \frac{F_i \phi_i(r)}{\omega_i^2 - \omega^2} \right|^2 dr d\omega \\
&\quad -2 \left(\frac{1}{2\pi} \int_{-\omega_c}^{\omega_c} \sum_{i=N+1}^{\infty} \frac{F_i k_i \Phi_i^2}{\omega_i^2 - \omega^2} d\omega \right) \\
&\quad + \frac{1}{2\pi} \int_{-\omega_c}^{\omega_c} \sum_{i=N+1}^{\infty} k_i^2 \Phi_i^2 d\omega,
\end{aligned}$$

where we have used the orthogonality property of the mode-shapes (4.2). The optimal set of parameters k_i^{opt}, $i = N+1, N+2, \ldots$ can now be determined via setting the derivatives of J with respect to k_i equal to zero, i.e. $\frac{\partial J}{\partial k_i} = 0$. We then find,

$$k_i^{opt} = \frac{1}{2\omega_c \Phi_i^2} \int_{-\omega_c}^{\omega_c} \frac{F_i \Phi_i^2}{\omega_i^2 - \omega^2} d\omega \tag{4.12}$$

which is equivalent to

$$k_i^{opt} = \frac{F_i}{2\omega_c\omega_i} \ln\left(\frac{\omega_i + \omega_c}{\omega_i - \omega_c}\right).$$

□

At this point, we intend to demonstrate how the K_i obtained from Theorem 4.1 is related to the value of K_i obtained from (4.4). For $x > 0$, the term $\ln(x)$ can be expanded as (see page 52 of [31]):

$$\ln(x) = 2\sum_{n=1}^{\infty} \frac{1}{2n-1}\left(\frac{x-1}{x+1}\right)^{(2n-1)}.$$

If we use the first term of the series to approximate $\ln\left(\frac{\omega_i+\omega_c}{\omega_i-\omega_c}\right)$ in (4.10), we obtain,

$$k_i = \frac{F_i}{\omega_i^2} \quad \text{for} \quad i = N+1, N+2, \ldots. \tag{4.13}$$

Hence, we recover (4.4). Therefore, the value of k_i suggested by (4.4) approximates the optimal k_i which minimizes (4.8).

To this end, we point out that the analysis given here ignores the effect of modal damping. There are two reasons for this. First, it is a difficult exercise to determine modal damping terms during the modeling phase (as discussed in Chapter 2). Second, every mode of a flexible structure is very lightly damped. Therefore, it is acceptable to first model the structure using the modal analysis technique, then use the approximation method explained above to correct the in-bandwidth dynamics of the model, and finally measure the damping associated with each in-bandwidth mode and add them to the corrected model. These dampings can then be easily embedded in the approximate model of the structure.

It should be noted that if the modal dampings are known for a large number of modes, it is readily possible to include the dampings in the model and carry out the optimization procedure explained above with some modifications. This approach, however, is not recommended. To find a three-mode approximation of a thirty-mode system, one would need to determine modal dampings for the first thirty modes. However, if the procedure discussed earlier is employed, one requires knowledge only of the first three modal dampings.

In Section 4.4.1, we will demonstrate how effective this approximation can be in capturing the effect of higher order modes on the frequency re-

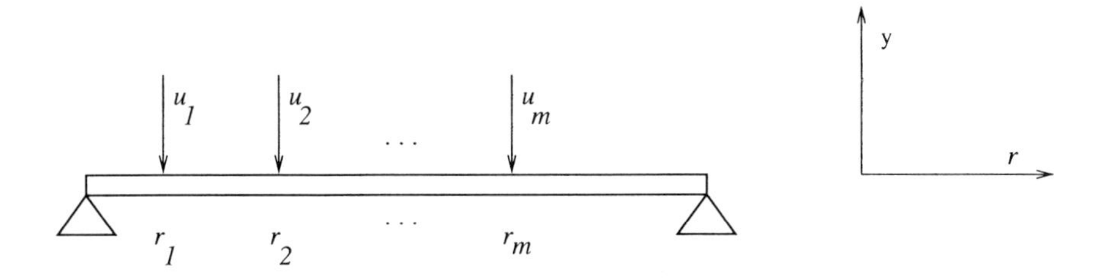

Fig. 4.1 A simply-supported beam with m point forces

sponse of the truncated system. But first, we extend the results of this section to the case of multi-input systems.

4.4 Extension to multi-input systems

In this section, we extend the procedure developed in Section 4.3, to the case of spatio-temporal systems that possess more than one input. The transfer function matrix is assumed to be of the form

$$G(s,r) = \sum_{i=1}^{\infty} \frac{\phi_i(r)}{s^2 + \omega_i^2} H_i. \tag{4.14}$$

Here,

$$H_i = \left[F_1^i \; F_2^i \; \ldots \; F_m^i \right],$$

where m is the number of actuators. Moreover, ϕ_is are assumed to satisfy similar orthogonality conditions, i.e., (4.2). For the simply-supported beam depicted in Figure 4.1, which is subject to m point forces at $r_1, \ldots, r_m$, this amounts to

$$F_s^i = \phi_i(r_s), \quad s = 1, 2, \ldots, m,$$

where $\phi_i(r)$ is given by

$$\phi_i(r) = \sqrt{\frac{2}{\rho AL}} \sin(\frac{i\pi r}{L}).$$

In parallel with the previous section, we approximate $G(s,r)$ with

$$\hat{G}(s,r) = G_N(s,r) + \Lambda(r), \tag{4.15}$$

where

$$\Lambda(r) = \sum_{i=N+1}^{\infty} \phi_i(r) K_i$$
$$K_i = \begin{bmatrix} k_1^i \; k_2^i \; \dots \; k_m^i \end{bmatrix}$$
$$G_N(s,r) = \sum_{i=1}^{N} \frac{\phi_i(r)}{s^2 + \omega_i^2} H_i.$$

Next, we consider the spatial cost function

$$J = \ll W(s,r)\left(G(s,r) - \hat{G}(s,r)\right) \gg_2^2, \tag{4.16}$$

where G and $\hat{G}$ are defined in (4.14) and (4.15) and W is defined in (4.9). The problem here is to determine the row vectors K_i, $i = N+1, N+2, \dots$ that would minimize the spatial cost function (4.16).

Theorem 4.2 *Consider the multi-input system (4.14) and its corresponding approximation (4.15). Then the row vectors K_i, $i = 1, 2, \dots$ that minimize (4.16) are given by:*

$$K_i = \frac{1}{2\omega_c \omega_i} \ln\left(\frac{\omega_i + \omega_c}{\omega_i - \omega_c}\right) H_i, \quad i = N+1, N+2, \dots . \tag{4.17}$$

Proof It can be verified that the spatial cost function J is equivalent to:

$$\begin{aligned} J = & \ll W(j\omega,r)\tilde{G}(j\omega,r) \gg_2^2 \\ & + \frac{1}{2\pi} \int_{-\infty}^{\infty} \int_{\mathcal{R}} \mathrm{tr}\{\Lambda(r)' W(j\omega,r)^* W(j\omega,r) \Lambda(r)\} dr d\omega \\ & -2 \times \frac{1}{2\pi} \int_{-\infty}^{\infty} \int_{\mathcal{R}} \left(\mathrm{tr}\{\tilde{G}(j\omega,r)^* W(j\omega,r)^* W(j\omega,r) \Lambda(r)\}\right) dr d\omega, \end{aligned}$$

where

$$\tilde{G}(s,r) = \sum_{i=N+1}^{\infty} \frac{\phi_i(r)}{s^2 + \omega_i^2} H_i.$$

This, in turn, can be re-written as:

$$J = \ll W(j\omega)\tilde{G}(j\omega, r) \gg_2^2$$
$$+\frac{1}{2\pi}\int_{-\omega_c}^{\omega_c}\int_{\mathcal{R}} \mathrm{tr}\{(\sum_{i=N+1}^{\infty} \phi_i(r)K_i')(\sum_{j=1}^{\infty} \phi_j(r)K_j)\}drd\omega$$
$$-2 \times \frac{1}{2\pi}\int_{-\omega_c}^{\omega_c}\int_{\mathcal{R}} \left(\mathrm{tr}\{(\sum_{i=N+1}^{\infty} \frac{\phi_i(r)}{\omega_i^2 - \omega^2} H_i')(\sum_{j=1}^{\infty} \phi_j(r)K_j)\} \right) drd\omega.$$

Applying the orthogonality conditions to this expression for J, we obtain

$$J = \ll W(j\omega)\tilde{G}(j\omega, r) \gg_2^2 + \frac{1}{2\pi}\int_{-\omega_c}^{\omega_c} \mathrm{tr}\{\sum_{i=N+1}^{\infty} \Phi_i^2 K_i' K_i\}d\omega$$
$$-2 \times \frac{1}{2\pi}\int_{-\omega_c}^{\omega_c} \mathrm{tr}\{\sum_{i=N+1}^{\infty} \frac{\Phi_i^2}{\omega_i^2 - \omega^2} H_i' K_i\}$$
$$= \ll W(j\omega)\tilde{G}(j\omega, r) \gg_2^2 + 2\omega_c \times \frac{1}{2\pi}\mathrm{tr}\{\sum_{i=N+1}^{\infty} \Phi_i^2 K_i' K_i\}$$
$$-2 \times \frac{1}{2\pi}\mathrm{tr}\{\sum_{i=N+1}^{\infty} \frac{\Phi_i^2}{\omega_i} \ln\left(\frac{\omega_i + \omega_c}{\omega_i - \omega_c}\right) H_i' K_i\}.$$

Differentiating J with respect to K_i (see page 592 of [58]), the optimum value of K_i is found to be:

$$K_i = \frac{1}{2\omega_c\omega_i} \ln\left(\frac{\omega_i + \omega_c}{\omega_i - \omega_c}\right) H_i.$$

□

Comparing the results of Theorems 4.11 and 4.2, we make the following observation.

Observation 4.4.1 Consider the multi-input system (4.14), and approximate each individual transfer function using the result of Theorem 4.1. Then the resulting transfer matrix is optimal in the sense of (4.16).

This is an interesting and non-trivial result, which is mainly due to the fact that all the individual transfer functions of (4.14) have similar poles.

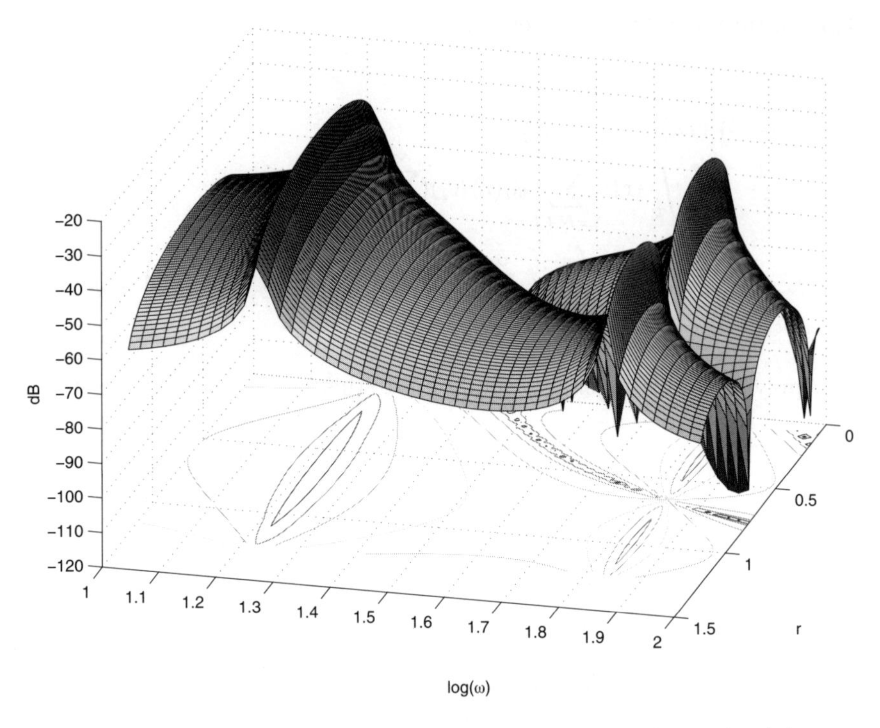

Fig. 4.2 Spatial frequency response of the beam from the point force input (N) to the displacement (m) based on a thirty-mode model

4.4.1 *Illustrative example*

In this section, we apply the approximation method which was developed in the previous section to a simply-supported beam model. The beam is shown in Figure 4.1, assuming that only the input u_1 is present. The parameters of the beam are given below,

$$\begin{aligned} L &= \text{Beam Length} = 1.3m, \\ r_1 &= 0.05m, \\ \rho A &= 0.6265kg/m \quad \text{and} \\ EI &= 5.329Nm^2. \end{aligned}$$

In this example, we are only interested in the first two modes of the beam. Figure 4.2 shows the spatial frequency response of the beam up to a frequency of 100 rad/sec. The model is obtained using the first thirty simply-supported modes of the beam as explained in Section 2.6.1. Fur-

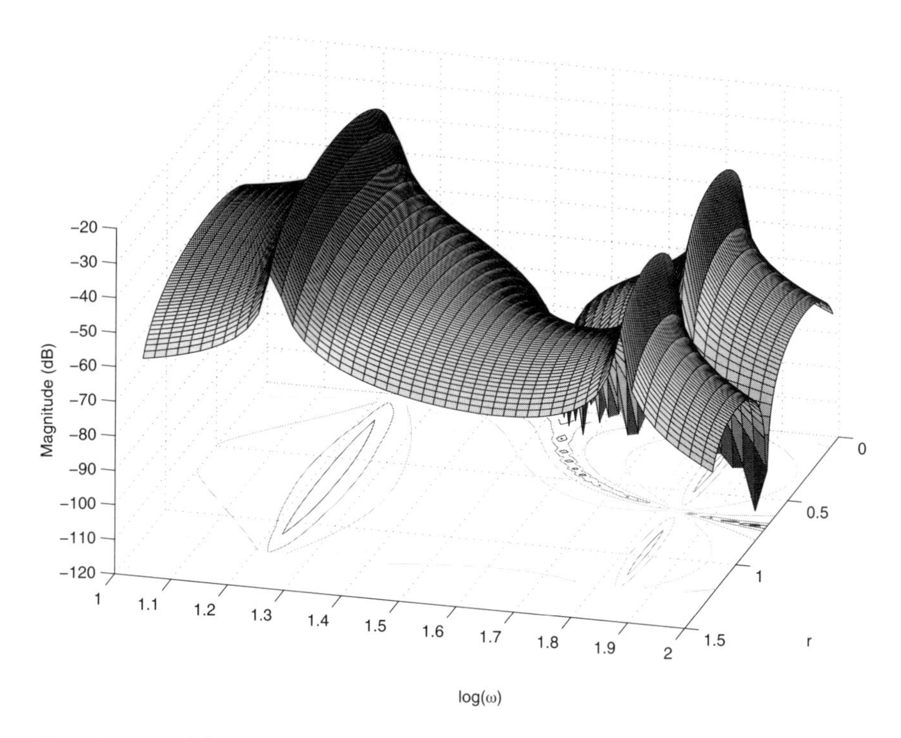

Fig. 4.3 Spatial frequency response of the beam from the point force input (N) to the displacement (m) based on a two-mode model

thermore, in the simulations, we add a damping of $\zeta_i = 0.003$ to each mode.

Figure 4.3 shows the spatial frequency response of the beam using only the first two modes. To have a clear picture of the spatial error caused by truncating the higher frequency modes, we plot the frequency response of the error system in Figure 4.4. Now, we approximate the truncated higher order modes by a spatial feed-through term as explained in the previous section. The spatial frequency response of this new system is plotted in Figure 4.5. Finally, we plot the spatial frequency response of the error system, i.e. the thirty-mode model and the two-mode model plus the correction term, in Figure 4.6. It can be observed that the approximation technique suggested here is a much better option than simply truncating the infinite-dimensional model, as is clear from Figures 4.4 and 4.6.

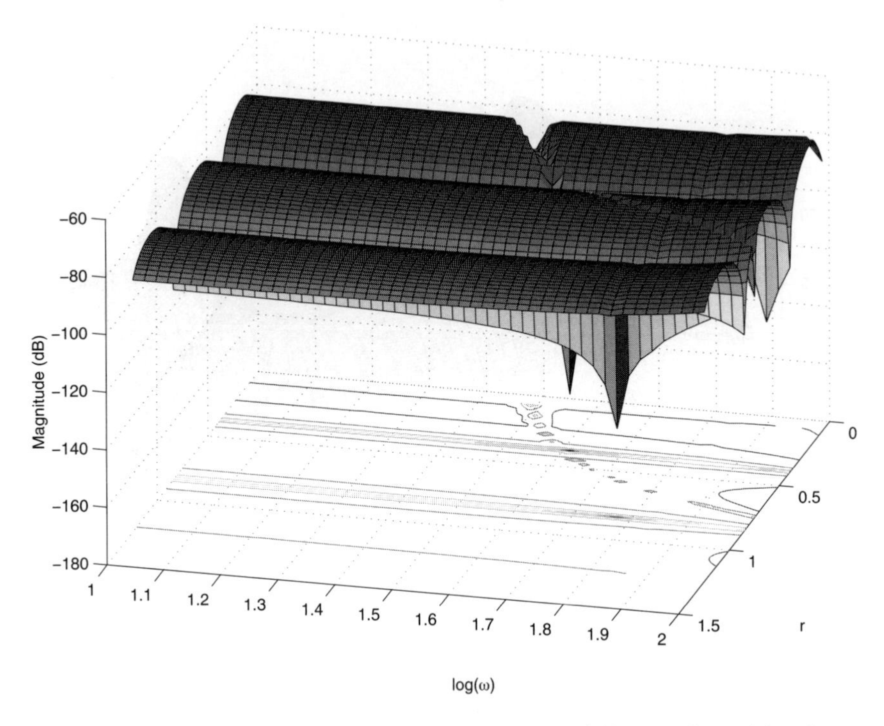

Fig. 4.4 Spatial frequency response of the error system (thirty-mode model and two-mode model)

4.5 Model correction using the spatial $\mathcal{H}_\infty$ norm

In this section we address a problem similar to the one solved in the previous section. However, now we use an alternative measure of performance, i.e. the spatial $\mathcal{H}_\infty$ norm of the error system. In the forthcoming chapters we will be using the spatial information embedded in the models of spatially distributed systems to design spatial H_2 and H_∞ controllers. This is our main motivation in developing model correction methods using the spatial H_2 and H_∞ norms.

We consider a multi-input system of the form (4.14) and its corresponding orthogonality condition (4.2). Our approach to reducing the truncation error is to add a feed-through term $K(r)$ to $G_N(s,r)$ such that the spatial $\mathcal{H}_\infty$ norm of the error system is minimized. Our model correction technique is based on approximating (4.14) with (4.15) such that the following cost function is minimized:

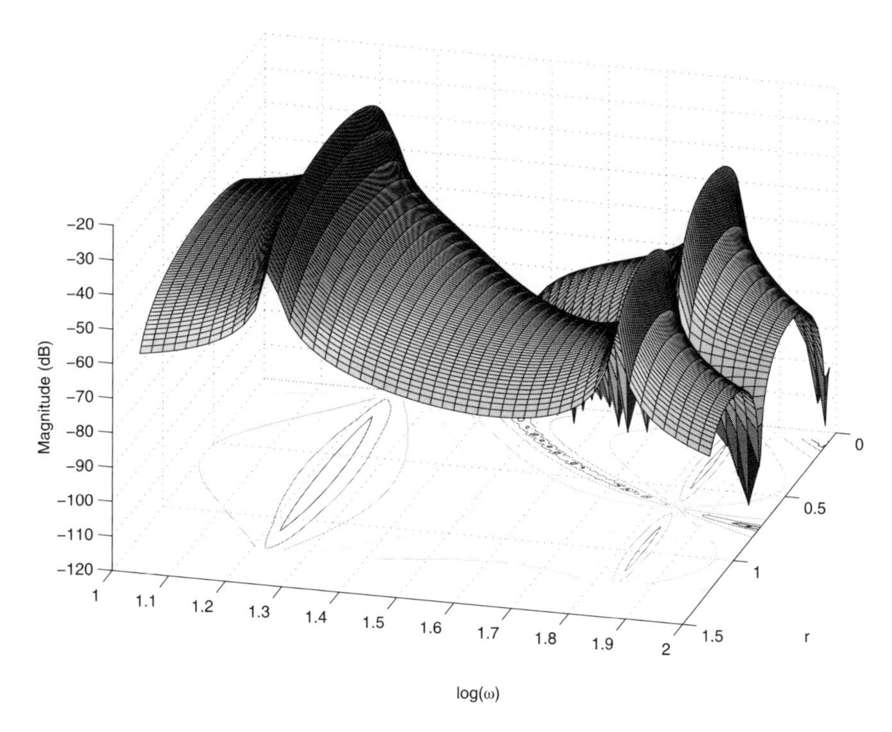

Fig. 4.5 Spatial frequency response of the beam based on two-mode model and a correction term

$$J = \ll W(s,r)\left(G(s,r) - \hat{G}(s,r)\right) \gg_{\infty}^{2}. \tag{4.18}$$

Here G and $\hat{G}$ are defined in (4.14) and (4.15), and $W(s,r)$ is an ideal low-pass weighting function distributed spatially over $\mathcal{R}$ with its cut-off frequency ω_c chosen to lie within the interval $\omega_c \in (\omega_N, \omega_{N+1})$ as in (4.9).

We notice that $G(s,r) - \hat{G}(s,r)$ has no resonant poles in the frequency range of $0 \leq \omega \leq \omega_c$. Therefore, the cost function (4.18) will be finite.

It turns out that the cost function (4.18) is minimized by setting

$$K(r) = \sum_{i=N+1}^{\infty} \phi_i(r) K_i^{opt} \tag{4.19}$$

with

$$K_i^{opt} = \frac{1}{2}\left(\frac{1}{\omega_i^2} + \frac{1}{\omega_i^2 - \omega_c^2}\right) H_i. \tag{4.20}$$

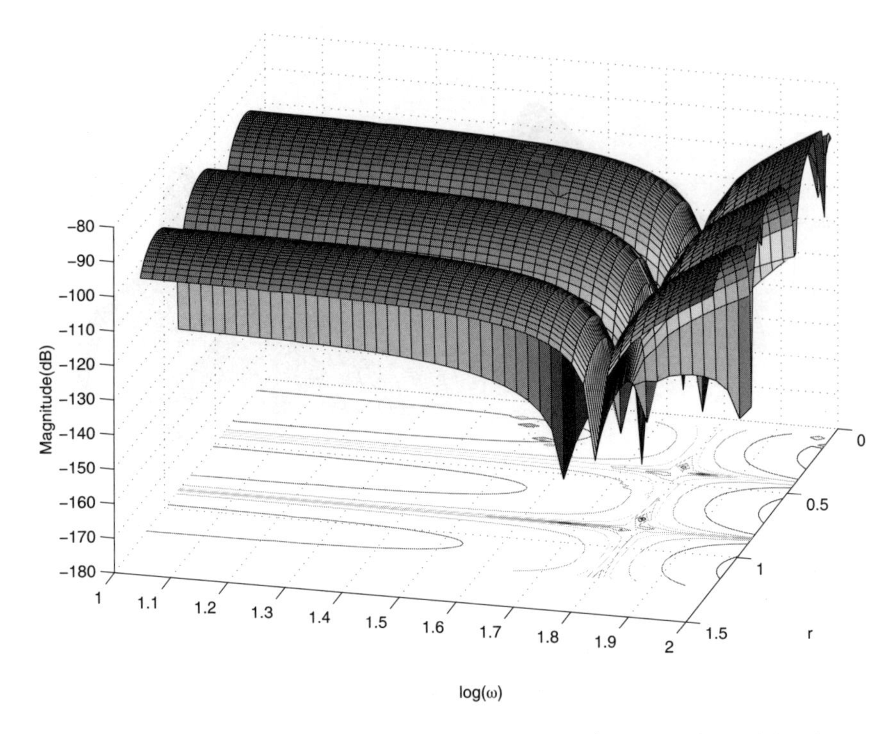

Fig. 4.6 Spatial frequency response of the error system (thirty-mode model and two-mode model plus the correction term)

To see this we first require two lemmas. In the first lemma we show that we can exploit orthogonality to express the cost function as the supremum over a closed frequency range of the maximum eigenvalue of a sum of error terms:

Lemma 4.1 *Define the $m \times m$ matrix*

$$E_i(\omega, K_i) = \Phi_i^2 \left[\frac{1}{\omega_i^2 - \omega^2} H_i - K_i\right]' \left[\frac{1}{\omega_i^2 - \omega^2} H_i - K_i\right]. \tag{4.21}$$

Then the cost function J in (4.18) can be written as

$$J = \sup_{-\omega_c \leq \omega \leq \omega_c} \lambda_{\max} \left[\sum_{i=N+1}^{\infty} E_i(\omega, K_i)\right]. \tag{4.22}$$

Proof Applying the definition of the spatial $\mathcal{H}_\infty$ norm (3.10) we may say

$$J = \sup_{\omega \in \mathbf{R}} \lambda_{\max} \left\{ \int_{\mathcal{R}} \begin{bmatrix} |W(j\omega, r)|^2 \left(G(j\omega, r) - \hat{G}(j\omega, r)\right)^* \\ \times \left(G(j\omega, r) - \hat{G}(j\omega, r)\right) \end{bmatrix} dr \right\}.$$

From the definition of the weighting function $W(j\omega, r)$ in (4.9) this reduces to

$$J = \sup_{-\omega_c \leq \omega \leq \omega_c} \lambda_{\max} \left\{ \int_{\mathcal{R}} \begin{bmatrix} \left(G(j\omega, r) - \hat{G}(j\omega, r)\right)^* \\ \times \left(G(j\omega, r) - \hat{G}(j\omega, r)\right) \end{bmatrix} dr \right\},$$

but

$$\begin{aligned}
&\int_{\mathcal{R}} \left(G(j\omega, r) - \hat{G}(j\omega, r)\right)^* \left(G(j\omega, r) - \hat{G}(j\omega, r)\right) dr \\
&= \int_{\mathcal{R}} \left(\sum_{i=N+1}^{\infty} \phi_i(r) \left[\frac{1}{\omega_i^2 - \omega^2} H_i - K_i \right] \right)' \\
&\qquad\qquad \times \left(\sum_{k=N+1}^{\infty} \phi_k(r) \left[\frac{1}{\omega_k^2 - \omega^2} H_k - K_k \right] \right) dr \\
&= \sum_{i=N+1}^{\infty} \Phi_i^2 \left[\frac{1}{\omega_i^2 - \omega^2} H_i - K_i \right]' \left[\frac{1}{\omega_i^2 - \omega^2} H_i - K_i \right] \\
&= \sum_{i=N+1}^{\infty} E_i(\omega, K_i),
\end{aligned}$$

hence the result. □

In our second lemma we consider the structure of each term $E_i(\omega, K_i)$. In particular we consider its structure over all frequencies $-\omega_c \leq \omega \leq \omega_c$ when we set

$$K_i = K_i^{opt}$$

with K_i^{opt} defined by (4.20). We also consider its structure over all row vectors K_i at frequencies $\omega = 0$ and $\omega = \omega_c$. An immediate corollary is that each row vector K_i^{opt} minimizes the cost function

$$J_i = \sup_{-\omega_c \leq \omega \leq \omega_c} \lambda_{\max} \left[E_i(\omega, K_i) \right].$$

Lemma 4.2 *With $E_i(\omega, K_i)$ defined by (4.21) and for K_i^{opt} defined by (4.20), then for any row vector X and for any K_i we can say*

(1) $E_i(0, K_i^{opt}) = E_i(\omega_c, K_i^{opt}) = \frac{1}{4}\Phi_i^2 (H_i' H_i) \left(\frac{1}{\omega_i^2} - \frac{1}{\omega_i^2 - \omega_c^2}\right)^2,$

(2) $XE_i(\omega, K_i^{opt})X' \leq XE_i(0, K_i^{opt})X'$ *for* $-\omega_c \leq \omega \leq \omega_c$ *and*

(3) $XE_i(0, K_i^{opt})X' \leq \max[XE_i(0, K_i)X', XE_i(\omega_c, K_i)X']$.

Proof

(1) This follows from simple calculation.

(2) We can evaluate

$$E_i(\omega, K_i^{opt}) = \frac{1}{4}\Phi_i^2 (H_i' H_i) \left(\frac{2}{\omega_i^2 - \omega^2} - \frac{1}{\omega_i^2} - \frac{1}{\omega_i^2 - \omega_c^2}\right)^2$$

so

$$\begin{aligned} &E_i(\omega, K_i^{opt}) - E_i(0, K_i^{opt}) \\ &= \frac{1}{4}\Phi_i^2 (H_i' H_i) \left[\left(\frac{2}{\omega_i^2 - \omega^2} - \frac{1}{\omega_i^2} - \frac{1}{\omega_i^2 - \omega_c^2}\right)^2 \right. \\ &\qquad\qquad \left. - \left(\frac{1}{\omega_i^2} - \frac{1}{\omega_i^2 - \omega_c^2}\right)^2\right] \\ &= \Phi_i^2 (H_i' H_i) \left(\frac{1}{\omega_i^2 - \omega^2} - \frac{1}{\omega_i^2}\right)\left(\frac{1}{\omega_i^2 - \omega^2} - \frac{1}{\omega_i^2 - \omega_c^2}\right). \end{aligned}$$

Hence

$$\begin{aligned} &XE_i(\omega, K_i^{opt})X' - XE_i(0, K_i^{opt})X' \\ &= \Phi_i^2 (XH_i')^2 \left(\frac{1}{\omega_i^2 - \omega^2} - \frac{1}{\omega_i^2}\right)\left(\frac{1}{\omega_i^2 - \omega^2} - \frac{1}{\omega_i^2 - \omega_c^2}\right) \\ &\leq 0 \end{aligned}$$

since the terms $\left(\frac{1}{\omega_i^2 - \omega^2} - \frac{1}{\omega_i^2}\right)$ and $\left(\frac{1}{\omega_i^2 - \omega^2} - \frac{1}{\omega_i^2 - \omega_c^2}\right)$ have opposite signs.

(3) By *reductio ad absurdum*:
Suppose we have both

$$XE_i(0, K_i)X' < XE_i(0, K_i^{opt})X'$$

and

$$XE_i(\omega_c, K_i)X' < XE_i(\omega_c, K_i^{opt})X'.$$

Evaluating the expressions we find both

$$\Phi_i^2 \left(\frac{1}{\omega_i^2} XH_i' - XK_i'\right)^2 - \Phi_i^2 \left(\frac{1}{\omega_i^2} XH_i' - XK_i^{opt'}\right)^2 < 0$$

and

$$\Phi_i^2 \left(\frac{1}{\omega_i^2 - \omega_c^2} X H_i' - X K_i' \right)^2 - \Phi_i^2 \left(\frac{1}{\omega_i^2 - \omega_c^2} X H_i' - X K_i^{opt\prime} \right)^2 < 0$$

and hence both

$$\Phi_i^2 \left(\frac{2}{\omega_i^2} X H_i' - X K_i' - X K_i^{opt\prime} \right) \left(X K_i^{opt\prime} - X K_i' \right) < 0$$

and

$$\Phi_i^2 \left(\frac{2}{\omega_i^2 + \omega_c^2} X H_i' - X K_i' - X K_i^{opt\prime} \right) \left(X K_i^{opt\prime} - X K_i' \right) < 0.$$

Adding we find

$$2\Phi_i^2 \left(\left[\frac{1}{\omega_i^2} + \frac{1}{\omega_i^2 + \omega_c^2} \right] X H_i' - X K_i' - X K_i^{opt\prime} \right) \times \left(X K_i^{opt\prime} - X K_i' \right) < 0.$$

Substituting for K_i^{opt} from (4.20) gives

$$2\Phi_i^2 \left(\frac{1}{2} \left[\frac{1}{\omega_i^2} + \frac{1}{\omega_i^2 + \omega_c^2} \right] X H_i' - X K_i' \right)^2 < 0$$

which cannot be true.

□

Corollary 4.1 *Each term K_i^{opt} defined by (4.20) satisfies:*

$$K_i^{opt} = \arg \inf_{K_i} \max_{-\omega_c \leq \omega \leq \omega_c} \lambda_{\max} \left[E_i(\omega, K_i) \right].$$

Proof The maximum eigenvalue is given by

$$\lambda_{\max} \left[E_i(\omega, K_i^{opt}) \right] = \max_{X \neq 0} \frac{X E_i(\omega, K_i^{opt}) X'}{X X'}$$

From part (ii) of Lemma 4.2 we see that

$$\lambda_{\max} \left[E_i(\omega, K_i^{opt}) \right] \leq \lambda_{\max} \left[E_i(0, K_i^{opt}) \right].$$

But similarly from part (iii)

$$\lambda_{\max}\left[E_i(0, K_i^{opt})\right] \leq \max\left\{\lambda_{\max}\left[E_i(0, K_i)\right], \lambda_{\max}\left[E_i(\omega_c, K_i)\right]\right\} \quad \text{for any } K_i$$
$$\leq \max_{-\omega_c \leq \omega \leq \omega_c} \lambda_{\max}\left[E_i(\omega, K_i)\right],$$

hence

$$\max_{-\omega_c \leq \omega \leq \omega_c} \lambda_{\max}\left[E_i(\omega, K_i^{opt})\right] \leq \max_{-\omega_c \leq \omega \leq \omega_c} \lambda_{\max}\left[E_i(\omega, K_i)\right] \text{ for all } K_i.$$

□

We are now ready to state our main result. Essentially we show that the cost function (4.18) has the same properties we showed for each term $E_i(\omega, K_i)$ in Lemma 4.2. The result (4.19) then follows as a corollary:

Theorem 4.3 *Define*

$$\bar{K}_M = \{K_{N+1}, K_{N+2}, \ldots, K_{N+M}\}$$

with

$$\bar{K}_M^{opt} = \{K_{N+1}^{opt}, K_{N+2}^{opt}, \ldots, K_{N+M}^{opt}\}.$$

Define also

$$S_M(\omega, \bar{K}_M) = \sum_{i=N+1}^{N+M} E_i(\omega, K_i).$$

Then for all M, for any row vector X and for any $\bar{K}_M$ we can say

(1)

$$S_M(0, \bar{K}_M^{opt}) = S_M(\omega_c, \bar{K}_M^{opt})$$
$$= \frac{1}{4}\sum_{i=N+1}^{N+M} \Phi_i^2\left(H_i'H_i\right)\left(\frac{1}{\omega_i^2} - \frac{1}{\omega_i^2 - \omega_c^2}\right)^2,$$

(2) $XS_M(\omega, \bar{K}_M^{opt})X' \leq XS_M(0, \bar{K}_M^{opt})X'$ *for* $-\omega_c \leq \omega \leq \omega_c$ *and*
(3) $XS_M(0, \bar{K}_M^{opt})X' \leq \max\left[XS_M(0, \bar{K}_M)X', XS_M(\omega_c, \bar{K}_M)X'\right]$.

Proof

(1) This follows immediately from Lemma 4.2.

(2) By induction:
Suppose the result is true for M. Then by supposition and from Lemma 4.2

$$\begin{aligned}
XS_{M+1}(\omega,\bar{K}_{M+1}^{opt})X' &= XS_M(\omega,\bar{K}_M^{opt})X' \\
&\quad + XE_{N+M+1}(\omega,K_{N+M+1}^{opt})X' \\
&\leq XS_M(0,\bar{K}_M^{opt})X' \\
&\quad + XE_{N+M+1}(0,K_{N+M+1}^{opt})X' \\
&= XS_{M+1}(0,\bar{K}_{M+1}^{opt})X'.
\end{aligned}$$

But we know from Lemma 4.2 that the result is true for $M = 1$.

(3) Following the same reasoning as for the proof of part (iii) of Lemma 4.2 we find that if both

$$XS_M(0,\bar{K}_M)X' < XE_i(0,\bar{K}_M^{opt})X'$$

and

$$XS_M(\omega_c,\bar{K}_M)X' < XS_M(0,\bar{K}_M^{opt})X',$$

then

$$\sum_{i=N+1}^{N+M} 2\Phi_i^2 \left(\frac{1}{2}\left[\frac{1}{\omega_i^2}+\frac{1}{\omega_i^2+\omega_c^2}\right]XH_i' - XK_i'\right)^2 < 0$$

which cannot be true.

□

Corollary 4.2 *The cost J in (4.18) is minimized by taking*

$$K(r) = \sum_{i=N+1}^{\infty} \phi_i(r)K_i^{opt}$$

Proof By the same reasoning as the Corollary to Lemma 4.2 we find

$$\bar{K}_M^{opt} = \arg \inf_{\bar{K}_M} \max_{-\omega_c\leq\omega\leq\omega_c} \lambda_{\max}\left[S_M(\omega,\bar{K}_M)\right].$$

Then letting $M \to \infty$ and applying Lemma 4.1 gives the result. □

Remark 4.1 *It can be observed from Corollary 4.2 that the optimal $K(r)$ consists of an infinite number of terms. In reality, however, it is enough to add up a finite, but perhaps large, number of terms to find K with a good precision. To see this, notice that as i increases, so does ω_i. Therefore, it should be clear from (4.20) that $K_i^{opt} \to 0$ as $i \to \infty$.*

Observation 4.5.1 A direct implication of Corollary (4.2) is that if our model correction technique is applied to each individual transfer function, the resulting corrected multi-input system will be optimal in the sense of (4.18). This is remarkable since the result can be applied to each transfer function term by term and similarly to each mode on a term by term basis.

4.5.1 *Illustrative example*

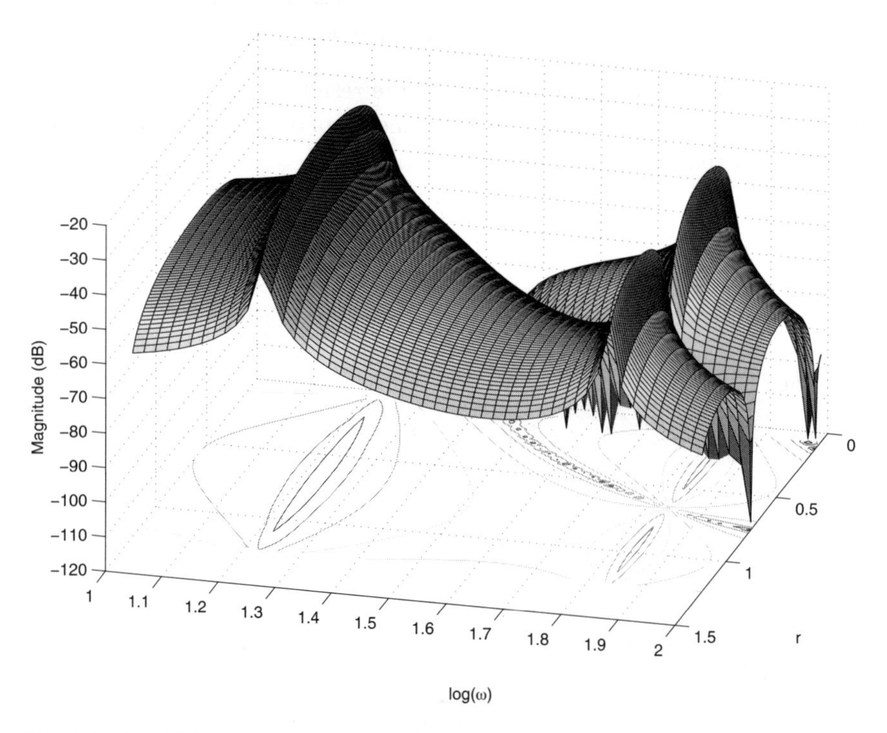

Fig. 4.7 Spatial frequency response of the beam from the point force input (N) to the displacement (m) based on two-mode model and a correction term

In this section, we apply the model correction method which was developed in the previous section to the simply-supported beam model depicted in Figure 3.2 whose physical parameters are given in Section 4.4.1. Moreover, in our simulations we allow for a damping ratio of 0.3% for all the modes.

In this example, we are only interested in the first two modes of the beam. Figure 4.2 shows the spatial frequency response of the beam up to a frequency of 100 rad/sec. The model is obtained using the first thirty

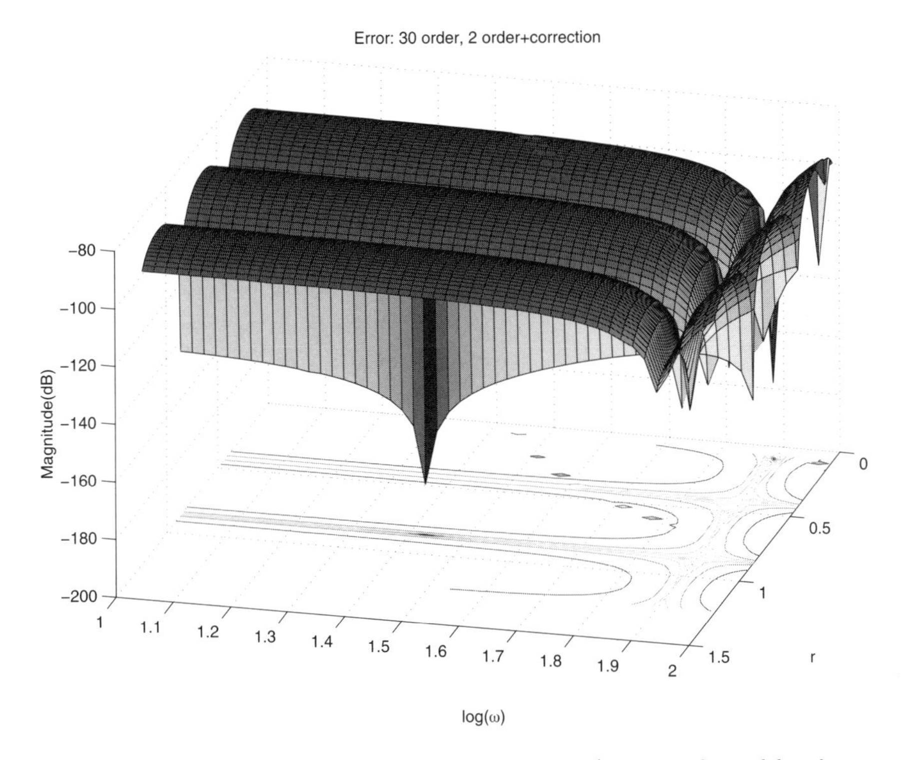

Fig. 4.8 Spatial frequency response of the error system (thirty-mode model and two-mode model plus the correction term)

pinned-pinned modes. Figure 4.3 shows the spatial frequency response of the beam using only the first two modes. Furthermore, the frequency response of the error system can be observed in Figure 4.4. Now, we approximate the truncated higher order modes by a spatial zero-frequency term as explained in the previous section. The spatial frequency response of this new system is plotted in Figure 4.7. Finally, we plot the spatial frequency response of the error system, i.e. the thirty-mode model and the two-mode model plus the correction term, in Figure 4.8. It can be observed that the approximation technique suggested in this paper is a much better option than simply truncating the model as is clear from Figures 4.4 and 4.8.

To make the implications of our model correction methodology clearer, we plot point-wise frequency response of the beam observed at four points along the structure. These plots are shown in Figure 4.9 and 4.10 and clearly demonstrate the effectiveness of the proposed technique.

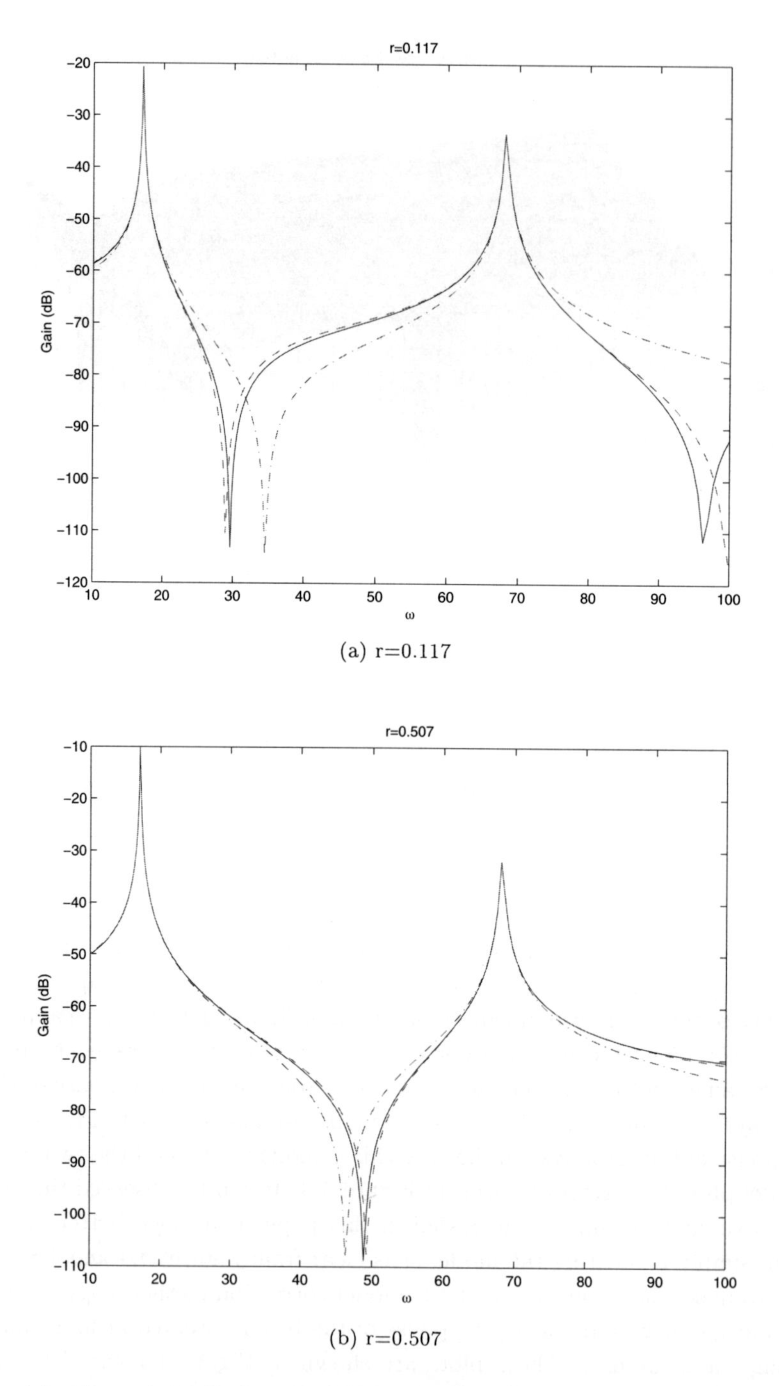

(a) r=0.117

(b) r=0.507

Fig. 4.9 Comparison of point-wise frequency responses of the beam: '–' thirty-mode model, '-.-' two-mode model, '- -' corrected two-mode model

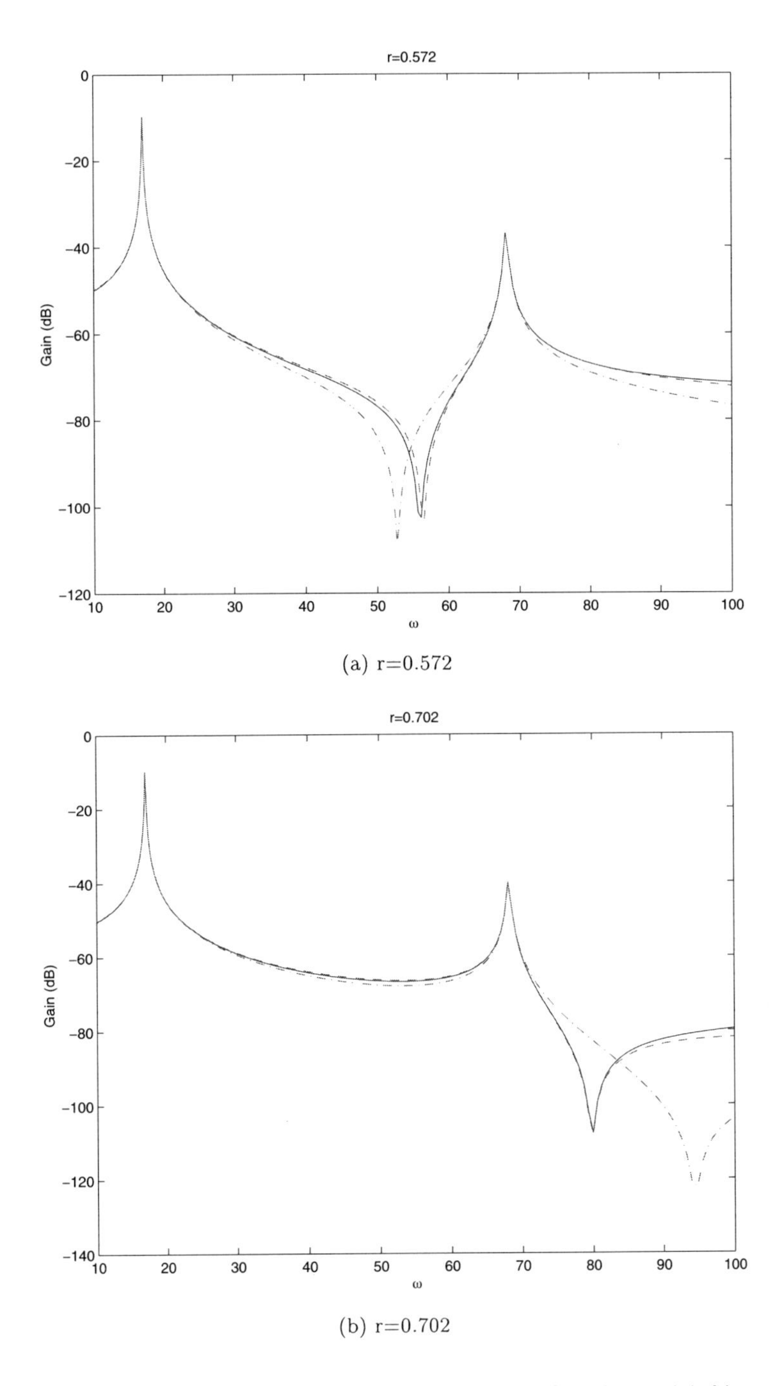

(a) r=0.572

(b) r=0.702

Fig. 4.10 Comparison of point-wise frequency responses of the beam: '–' thirty-mode model, '-.-' two-mode model, '- -' corrected two-mode model

4.6 Model correction for point-wise models of structures

So far in this chapter we have studied spatially distributed models of structures. Specifically, we have addressed the problem of compensating for the truncation error in a spatially distributed setting. In the forthcoming chapters we will see that such models are needed in designing high-performance feedback controllers with a view to minimizing structural vibration over the entire body of flexible structures.

This section addresses another important problem, that of model correction for point-wise models of structures. In a typical feedback control problem, a number of sensors and actuators are attached to the structure for the purposes of sensing and actuation. Typical actuators are shakers and piezoelectric transducers while a typical sensor may be an accelerometer, a strain gage, or even another piezoelectric transducer. The resulting system will have a finite number of inputs and a finite number of outputs. However, the underlying distributed parameter nature of the system will imply that its model will consist of an infinite number of modes. Subsequently, to simplify such a model, it will be truncated. In this section we will study the adverse effect of truncation on point-wise models of structures and how it may be compensated for.

4.6.1 *Model correction for SISO models*

Let us consider the model of a structure represented by:

$$G(s) = \sum_{i=1}^{\infty} \frac{F_i}{s^2 + \omega_i^2}. \tag{4.23}$$

We point out that a more precise version of (4.23) can be written as

$$G(s) = \sum_{i=1}^{\infty} \frac{F_i}{s^2 + 2\zeta_i s + \omega_i^2}.$$

However, as we argued earlier, we choose to ignore the effect of modal dampings at this stage of the analysis. The model (4.23) is then truncated by keeping the first N modes of the system that lie within the bandwidth of interest and removing the mode $N + 1$ and above, i.e.

$$G_N(s) = \sum_{i=1}^{N} \frac{F_i}{s^2 + \omega_i^2}. \tag{4.24}$$

By removing the high-frequency modes, we introduce an error that manifests itself by perturbing the zeros of $G_N(s)$, i.e., although the in-bandwidth poles of $G(s)$ are similar to the poles of $G_N(s)$, the zeros of $G_N(s)$ may be significantly displaced. It should be clear by now that this effect is mainly due to the fact that by removing modes $N+1$ and higher, we have removed their DC content as well. This effect can be worse, if the senor and actuator are collocated since in this case all the parameters F_i, are positive [11]. Hence, all the truncated modes are in phase at DC. As explained earlier, this problem can be rectified, to some extent, by adding a zero frequency term to $G_N(s)$. That is,

$$\hat{G}(s) = G_N(s) + K, \tag{4.25}$$

where

$$K = \sum_{i=N+1}^{\infty} \frac{F_i}{\omega_i^2}.$$

The reason behind this choice of K is that at lower frequencies one can ignore the effect of dynamical responses of higher order modes since they are much smaller than the forced responses at those frequencies.

In this section, we will attempt to determine a value for K such that the following cost function is minimized,

$$\|(G(s) - \hat{G}(s))W(s)\|_2^2. \tag{4.26}$$

Here, $\|f(s)\|_2^2 = \frac{1}{2\pi}\int_{-\infty}^{\infty} |f(j\omega)|^2 d\omega$ and $G(s)$ and $\hat{G}(s)$ are defined as in (4.23) and (4.25). Furthermore, $W(s)$ is an ideal low-pass weighting function with its cut-off frequency ω_c chosen to lie within the interval $\omega_c \in (\omega_N, \omega_{N+1})$. That is,

$$|W(j\omega)| = \begin{cases} 1 & \text{for } \; -\omega_c \leq \omega \leq \omega_c \\ 0 & \text{elsewhere.} \end{cases}$$

This choice of W will ensure that the cost function (4.26) will remain finite.

To this end, it should be clear that a K chosen to minimize (4.26) will minimize the effect of out-of-bandwidth dynamics of $G(s)$ on $\hat{G}(s)$ in an $\mathcal{H}_2$ optimal sense. Notice that the cost function (4.26) conveys no information at frequencies higher than ω_c.

It can be verified that (4.26) is equivalent to

$$\left\| \left(\sum_{i=N+1}^{\infty} \frac{F_i}{s^2+\omega_i^2} - K \right) W(s) \right\|_2^2 . \tag{4.27}$$

Let

$$\tilde{G}(s) = \sum_{i=N+1}^{\infty} \frac{F_i}{s^2+\omega_i^2}.$$

It is straightforward to show that (4.27) is equivalent to

$$\|\tilde{G}W\|_2^2 + K^2\|W\|_2^2 - K(<\tilde{G}W, W> + <W, \tilde{G}W>) \tag{4.28}$$

where $<f,g>= \frac{1}{2\pi}\int_{-\infty}^{\infty} f^*(j\omega)g(j\omega)d\omega$. It can be verified that the K that minimizes (4.28) is given by

$$K = \frac{<\tilde{G}W, W> + <W, \tilde{G}W>}{2\|W\|_2^2} \tag{4.29}$$

$$= \frac{\int_{-\infty}^{\infty} \mathbf{Re}(\tilde{G}(j\omega))|W(j\omega)|^2 d\omega}{\int_{-\infty}^{\infty} |W(j\omega)|^2 d\omega} \tag{4.30}$$

$$= \frac{\int_{-\infty}^{\infty} (\sum_{i=N+1}^{\infty} \frac{F_i}{\omega_i^2-\omega^2})|W(j\omega)|^2 d\omega}{\int_{-\infty}^{\infty} |W(j\omega)|^2 d\omega} \tag{4.31}$$

where $\mathbf{Re}(f)$ represents the real part of the complex number f. Hence, to obtain the optimal K, one has to carry out the following integration.

$$K = \frac{1}{2\omega_c}\int_{-\omega_c}^{\omega_c} \sum_{i=N+1}^{\infty} \frac{F_i}{\omega_i^2-\omega^2} d\omega. \tag{4.32}$$

The optimal value of K is then found to be

$$K_{opt} = \frac{1}{2\omega_c} \sum_{i=N+1}^{\infty} \frac{F_i}{\omega_i} \ln\left(\frac{\omega_i+\omega_c}{\omega_i-\omega_c}\right). \tag{4.33}$$

4.7 Extension to multi-variable point-wise systems

Next, we extend our model correction technique to multi-variable transfer functions. This is an important issue since in many cases it may not be practical to achieve the required performance by a single actuator and sensor. If several actuators and sensors are to be used, and the multi-variable model is to be truncated, it is essential to capture the effect of higher order modes on the remaining in-bandwidth modes, as we did in the SISO case. In the multi-variable case, the transfer function matrix of the system is given by:

$$G(s) = \sum_{i=1}^{\infty} \frac{\Psi_i}{s^2 + \omega_i^2}, \tag{4.34}$$

where, $\Psi_i \in \mathbf{R}^{m \times n}$, i.e. m sensors and n actuators. The multi-variable transfer function matrix $G(s)$ has a very interesting property as all of its individual transfer functions share similar poles. Moreover, if $m = n$ and all the actuators and sensors are collocated, the diagonal transfer functions will possess minimum-phase zeros only. However, the off-diagonal transfer functions may have non-minimum-phase zeros since they correspond to non-collocated actuators and sensors.

It is our intention to approximate $G(s)$ by a finite number of modes, say N modes only. In this case, however, we choose to approximate the effect of higher order modes on the low-frequency dynamics of $G(s)$ by a constant matrix. That is, we approximate (4.34) by

$$\hat{G}(s) = \sum_{i=1}^{N} \frac{\Psi_i}{s^2 + \omega_i^2} + K. \tag{4.35}$$

We will determine K such that the following cost function is minimized:

$$J = \left\| W(s)(G(s) - \hat{G}(s)) \right\|_2^2, \tag{4.36}$$

where for a multi-variable F, $\|F(s)\|_2^2 = \frac{1}{2\pi} \int_{-\infty}^{\infty} \mathrm{tr}\{F^*(j\omega)F(j\omega)\} d\omega$. Here, W is chosen to be a diagonal matrix, where the diagonal elements are ideal low-pass filters $W = diag(w, w, \ldots, w)$ and w is an ideal low-pass filter as described earlier in this chapter. The cost function (4.36) can be re-written as

$$J = \|W(s)(\tilde{G}(s) - K)\|_2^2,$$

where

$$\tilde{G}(s) = \sum_{i=N+1}^{\infty} \frac{\Psi_i}{s^2 + \omega_i^2}.$$

Therefore,

$$J = \|W\tilde{G}\|_2^2 + \|WK\|_2^2 - (< W\tilde{G}, WK > + < WK, W\tilde{G} >),$$

where $< F, G >= \frac{1}{2\pi}\int_{-\infty}^{\infty} \mathrm{tr}\{F^*(j\omega)G(j\omega)\}d\omega$. The cost function can then be written as:

$$J = \|W\tilde{G}\|_2^2 + \frac{1}{2\pi}\int_{-\infty}^{\infty} \mathrm{tr}\{K'W(j\omega)^*W(j\omega)K\}d\omega$$
$$-\frac{1}{2\pi}\int_{-\infty}^{\infty} \begin{bmatrix} \mathrm{tr}\{\tilde{G}(j\omega)^*W(j\omega)^*W(j\omega)K\} \\ +\mathrm{tr}\{K'W(j\omega)^*W(j\omega)\tilde{G}(j\omega)\} \end{bmatrix} d\omega.$$

Differentiating J with respect to K (see page 592 of [58]), we obtain the optimum value of K.

$$\begin{aligned} K_{opt} &= \left(\int_{-\infty}^{\infty} W(j\omega)^*W(j\omega)d\omega\right)^{-1} \\ &\quad \times \left(\int_{-\infty}^{\infty} W(j\omega)^*W(j\omega)\mathbf{Re}\{\tilde{G}(j\omega)\}d\omega\right) \\ &= \frac{1}{2\omega_c}\int_{-\omega_c}^{\omega_c} \mathbf{Re}\{\tilde{G}(j\omega)\}d\omega \\ &= \frac{1}{2\omega_c}\int_{-\omega_c}^{\omega_c} \sum_{i=N+1}^{\infty} \frac{\Psi_i}{\omega_i^2 - \omega^2} d\omega \\ &= \frac{1}{2\omega_c}\sum_{i=N+1}^{\infty} \frac{1}{\omega_i} \ln\left(\frac{\omega_i + \omega_c}{\omega_i - \omega_c}\right)\Psi_i. \end{aligned}$$

Observation 4.7.1 An implication of this result is that if K_{opt} determined in (4.33) is used to approximate the effect of out-of-bandwidth modes on the individual truncated transfer functions of (4.34), the obtained multivariable transfer matrix will be optimal in the sense of (4.36).

Table 4.1 Parameters of the beam apparatus

Parameter	Value
Beam Length, L	0.6 m
Beam width	0.05 m
Beam thickness, t_b	0.003 m
Young's Modulus, E_b	$65 \times 10^9 N/m^2$ m
Density, ρ	$2650 kg/m^3$ m
Piezoceramic position, r_1	0.05 m
Piezoceramic position, r_2	0.12 m

4.8 Model correction for a piezoelectric laminate beam

In this section we apply the model correction method developed in the previous section to the model of a simply-supported piezoelectric laminate beam. The experimental beam apparatus, as illustrated in Figure 4.11, consists of a uniform aluminum beam of rectangular cross section. The beam parameters are given in Table 4.1.

A pair of piezoelectric ceramic patches are bonded symmetrically to either side of the beam structure, at 0.05 m from one end of the beam. The piezoceramic elements used on the experimental structure are PIC151 patches. The important physical parameters for the PIC151 piezoelectric patches are given in Table 4.2. Tiny screws are used to attach the two ends of the beam to a pair of shims and the shims are clamped at both ends. The resulting boundary conditions at the two ends of the beam are very close to pinned boundary conditions. The clamps are fastened to a secure structure, i.e. a Newport RS3000 optical table with passive isolation structural supports. The optical table is used to lower ambient noise levels, such as building transmission vibrations, which may effect experimental measurements.

The experiment is depicted in Figure 4.12. A Hewlett Packard model 89410A vector analyzer is used to determine the frequency response of the piezoelectric laminate beam. As explained in reference [1], the transfer function of the laminate may have to be modified to allow for the effect of finite input resistance of the measurement device. If the total resistance of the measuring device is R_m and the total capacitance of the sensing piezoelectric patch is C_k, in the transfer function (2.121), Ω_k should be replaced with

$$\Omega_k \left(\frac{sR_m C_k}{sR_m C_k + 1} \right).$$

The input resistance of HP89410A is 1 $M\Omega$. To reduce the effect of this

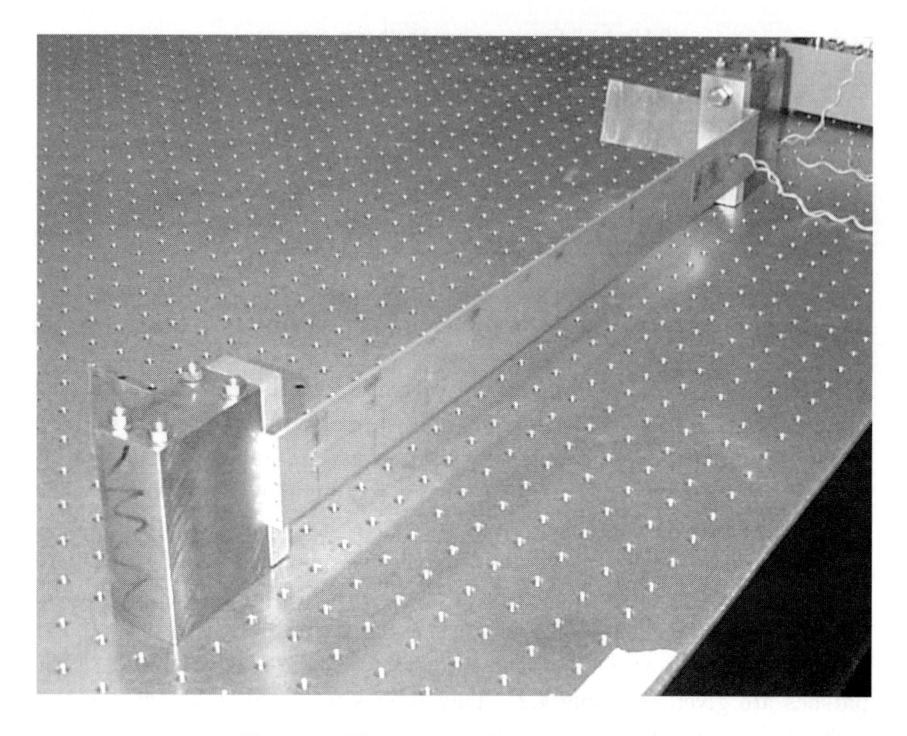

Fig. 4.11 The piezoelectric laminate beam

Table 4.2 Physical parameters of the PIC151 patches

Charge Constant, d_{31}	-210×10^{-12} m/v
Voltage Constant, g_{31}	-11.5×10^{-3} Vm/N
Coupling Coefficient	0.340
Capacitance, C_k	104.82 nF
Piezoceramic width	0.025 m
Piezoceramic thickness t_a	0.25×10^{-3} m

low input resistance on our measurement, a Tektronix P6201 active probe with 1 $M\Omega$ input resistance is used. This means that the high-pass cut-off frequency $f_c = \frac{1}{R_m C_k}$ is moved below 5 Hz which is quite acceptable for our purposes.

A model of the beam is determined using equation (2.121). Here, the mode shapes and resonance frequencies are determined as explained in Section 2. Our model consists of the first 150 modes of the beam. This is enough to give us an accurate model of the structure in the frequency range of up to 500 Hz. There are five modes within this particular band-

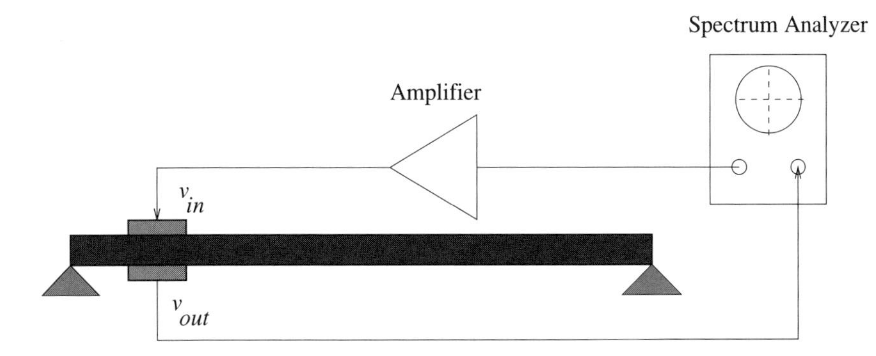

Fig. 4.12 The experimental set-up

Table 4.3 The first eight resonance frequencies of the simply-supported beam

Mode	f_i (Hz)	Mode	f_i (Hz)
1	21.3	5	466.63
2	77.13	6	670.59
3	169.92	7	911.63
4	299.74	8	1189.7

width. Hence, we are interested in working with a five-mode model of the structure.

In Figure 4.13 we compare the frequency response of the beam based on its first 150 modes with our experimental measurements in up to 500 Hz range. It can be observed that the two models are very close. In Figure 4.14 we plot the experimental data and the frequency response of the five-mode model. This figure clearly demonstrates the error that is introduced by truncation. In Figure 4.15 we plot our experimental data and the frequency response of the fifty-mode model of the beam. Some improvement can be observed, however, the difference between the frequency responses of the two systems is still unacceptably high. Now, we approximate the effect of truncated modes on the five-mode model of the system via the equation (4.33). The frequency ω_c is chosen to be $(\omega_5 + \omega_6)/2$, i.e. 3572.7 rad/sec. In Figure 4.16, we plot and compare the frequency response of the corrected five-mode model of the beam with our experimental measurements. It can be observed that the difference between the two frequency responses in the frequency range of interest is minimal.

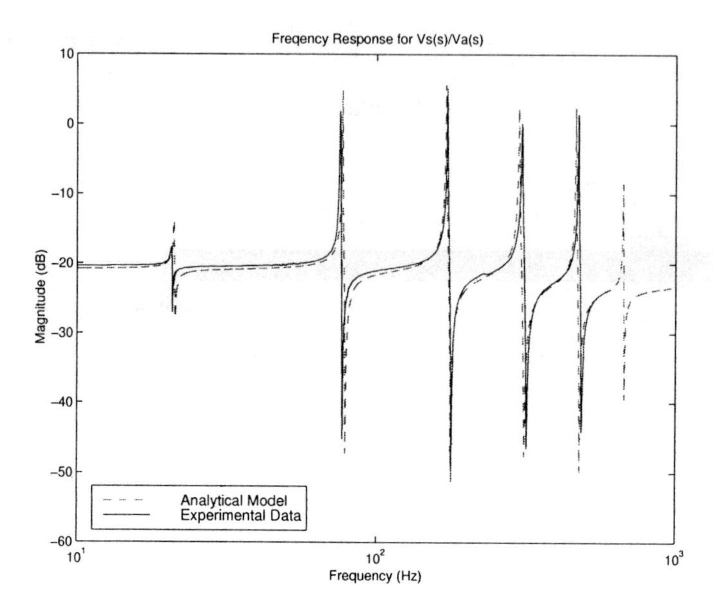

Fig. 4.13 Comparison of the frequency responses of the 150-mode model of the laminate with the experimental data

4.9 Conclusions

In this chapter we investigated the effect of truncation on the models of spatially distributed systems introduced in Chapter 2. We observed that truncating the model could perturb zeros of the system. Precise knowledge of zeros is important in feedback control systems. We demonstrated that by adding a feed-through term to the truncated model, the location of system zeros could be significantly corrected. Finally, we investigated this effect by performing a number of experiments on a simply-supported piezoelectric laminate beam.

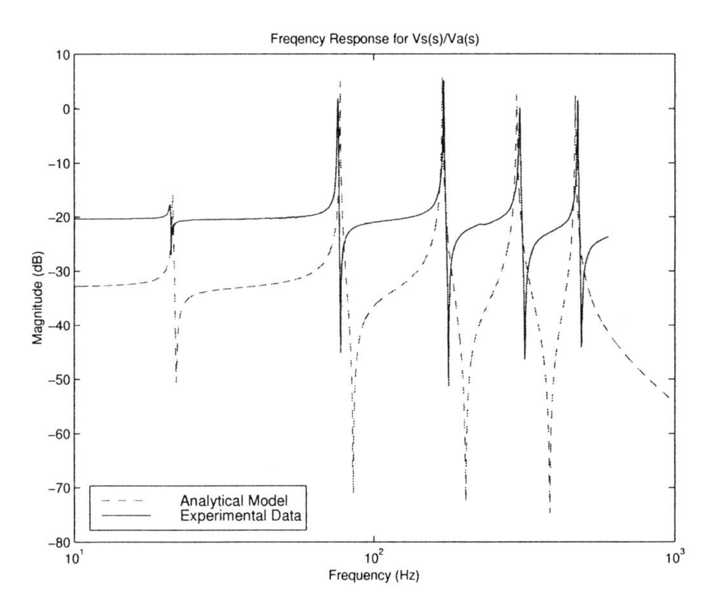

Fig. 4.14 Comparison of the frequency responses of the 5-mode model of the laminate with the measured frequency response

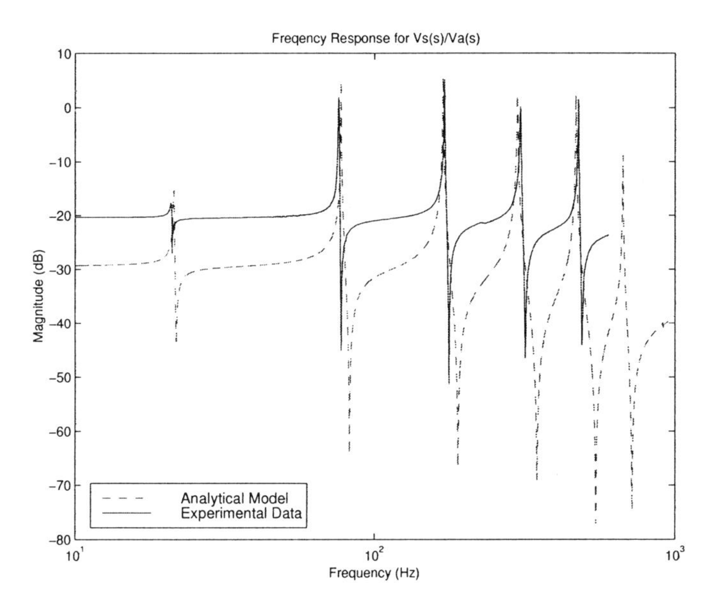

Fig. 4.15 Comparison of the frequency responses of the 50-mode model of the laminate with the measured frequency response

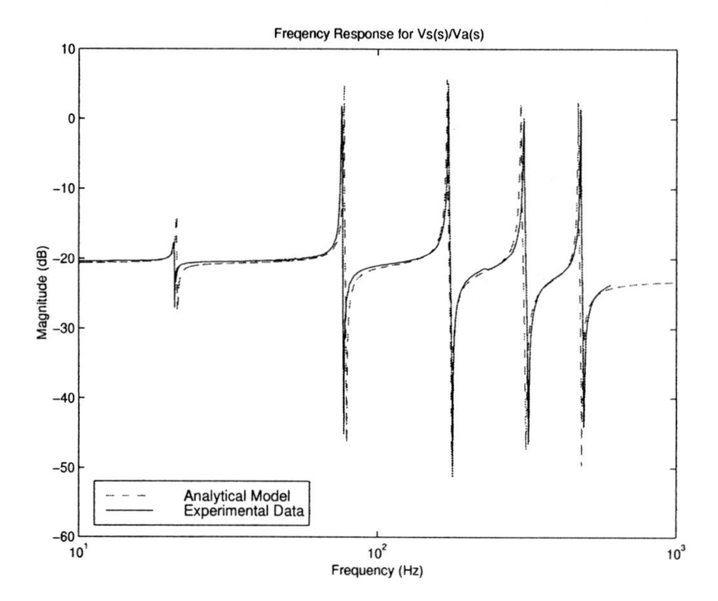

Fig. 4.16 Comparison of the frequency responses of the corrected model of the laminate with the measured frequency response

Chapter 5

Spatial Control

5.1 Introduction

In this chapter we study the problem of spatial control for a specific class of spatially distributed systems. The problem is relevant and well defined for a number of systems including those introduced in Chapter 2.

A difficulty that may arise in vibration control of flexible structures is that if the controller is designed with a view to minimizing vibration at a specific point, or a limited number of points, the resulting controller may not perform well at other points on the surface of the structure. Hence, the use of standard design tools, such as $\mathcal{H}_\infty$ or $\mathcal{H}_2$ control methodologies, may not result in effective reduction of vibrations over the entire body of the structure. The cost functions minimized during the design phase may convey little or no information on the spatial nature of the system. This issue is addressed by redefining the $\mathcal{H}_\infty$ and $\mathcal{H}_2$ [112, 32] control problems in a meaningful way.

Understanding the spatial structure of the class of systems described in Chapter 2 along with the model reduction and model correction procedures studied in Chapters 3 and 4, enables us to define optimal control problems in which the cost functions contain considerable information on the spatial dynamics of the systems. Furthermore, these problems can be solved by converting them to equivalent optimal control problems that can be subsequently solved via standard software.

In this chapter the concept of spatial control is applied to a very specific problem. That is, vibration control of a flexible beam using piezoelectric actuators and sensors. The proposed methodology, however, is applicable to a large class of systems. These are spatially distributed linear time-invariant systems with point-wise actuators and sensors. Active noise control in acoustic enclosures [81] using loudspeakers and microphones is another example of problems where similar techniques can be applied. In particular, there have been recent developments in understanding performance limitations associated with active noise control systems which strongly hints that spatial control may be a viable proposition for such systems [104, 105].

5.2 Spatial $\mathcal{H}_\infty$ control problem

In this section we consider the problem of the spatial $\mathcal{H}_\infty$ control for a spatially distributed linear time-invariant system. The system is assumed to be of the form

$$\begin{aligned} \dot{x}(t) &= Ax(t) + B_1 w(t) + B_2 u(t) \\ z(t,r) &= C_1(r)x(t) + D_{11}(r)w(t) + D_{12}(r)u(t) \\ y(t) &= C_2 x(t) + D_{21} w(t) + D_{22} u(t), \end{aligned} \tag{5.1}$$

where $r \in \mathcal{R}$ is the spatial coordinate, $x \in \mathbf{R}^n$ is the state of the system, $w \in \mathbf{R}^p$ is the disturbance input, $u \in \mathbf{R}^m$ is the control input, $z \in \mathbf{R}^q \times \mathcal{R}$ is the performance output and $y \in \mathbf{R}^\ell$ is the measured output.

The spatial $\mathcal{H}_\infty$ control problem is to design a controller

$$\begin{aligned} \dot{x}_k(t) &= A_k x_k(t) + B_k y(t) \\ u(t) &= C_k x_k(t) + D_k y(t) \end{aligned} \tag{5.2}$$

such that the closed-loop system satisfies

$$\inf_{K \in U} \sup_{w \in \mathcal{L}_2[0,\infty)} J_\infty < \gamma^2, \tag{5.3}$$

where U is the set of all stabilizing controllers and

$$J_\infty = \frac{\int_0^\infty \int_{\mathcal{R}} z(t,r)'Q(r)z(t,r)drdt}{\int_0^\infty w(t)'w(t)dt} \tag{5.4}$$

Here, $Q(r)$ is a spatial weighting function. The purpose of $Q(r)$ is to emphasize the region over which the effect of the disturbance is to be reduced more substantially. The numerator in (5.4) is the weighted spatial $\mathcal{H}_2$ norm of the performance signal $z(t,r)$. Therefore, J_∞ can be interpreted as the ratio of the energy of the spatially distributed signal to that of the disturbance signal, w. The control problem is depicted in Figure 5.1.

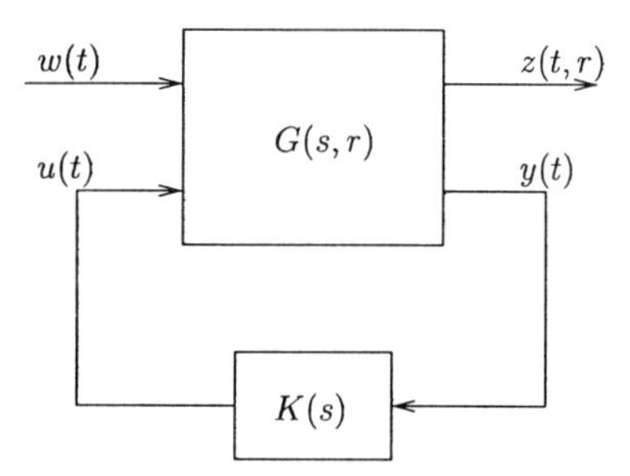

Fig. 5.1 Block diagram representation of the spatial $\mathcal{H}_\infty$ control problem

This problem can be shown to be equivalent to an ordinary $\mathcal{H}_\infty$ control problem by observing that

$$\begin{aligned} J_\infty &= \frac{\int_0^\infty \int_{\mathcal{R}} z(t,r)'Q(r)z(t,r)drdt}{\int_0^\infty w(t)'w(t)dt} \\ &= \frac{\int_0^\infty \tilde{z}(t)'\tilde{z}(t)dt}{\int_0^\infty w(t)'w(t)dt}, \end{aligned}$$

where

$$\tilde{z}(t) = \Pi x(t) + \Theta_{11}w(t) + \Theta_{12}u(t).$$

Here, matrices Π, Θ_{11} and Θ_{12} are determined from

$$\Gamma = \left[\Pi \; \Theta_{11} \; \Theta_{12}\right],$$

where

$$\Gamma'\Gamma = \int_{\mathcal{R}} \begin{bmatrix} C(r)' \\ D_{11}(r)' \\ D_{12}(r)' \end{bmatrix} Q(r) \left[C(r) \; D_{11}(r) \; D_{12}(r)\right] dr. \tag{5.5}$$

In a general setting, the spatial integral in (5.5) may have to be determined numerically. However, in certain cases, a simple solution can be found. For instance, if (5.1) is associated with a flexible structure, then the matrices $C_1(r)$, $D_{11}(r)$ and $D_{12}(r)$ will consist of mode shapes of the system which may satisfy certain orthogonality conditions. In such a case, the integral in (5.5) can be evaluated easily, as long as $Q(r) = 1$.

Choice of the spatial weighting function $Q(r)$ in (5.1) depends on the specific performance requirements determined by the designer. As a rule of thumb, $Q(r)$ should be large around the regions in $\mathcal{R}$ where the effect of the input disturbance is to be reduced more heavily. An extreme case is when the objective is to reduce the effect of disturbance in a number of discrete points, e.g., $\{r_1, r_2, \ldots, r_k\} \in \mathcal{R}$. Then we may choose

$$Q(r) = \sum_{i=1}^{k} \delta(r - r_i), \tag{5.6}$$

where $\delta(r - r_i)$ is the Dirac delta function at the point r_i. In this case, it can be verified that the spatial $\mathcal{H}_\infty$ control problem (5.1), (5.4) reduces to the problem defined by the system:

$$\begin{aligned}
\dot{x}(t) &= Ax(t) + B_1 w(t) + B_2 u(t) \\
\hat{z}(t) &= \hat{C}_1 x(t) + \hat{D}_{11} w(t) + \hat{D}_{12} u(t) \\
y(t) &= C_2 x(t) + D_{21} w(t) + D_{22} u(t)
\end{aligned} \tag{5.7}$$

and the performance objective

$$J_\infty = \frac{\int_0^\infty \hat{z}(t)'\hat{z}(t)dt}{\int_0^\infty w(t)'w(t)dt} \tag{5.8}$$

in which

$$\begin{aligned}
\hat{C}_1 &= [C_1(r_1)' \; C_1(r_2)' \; \ldots \; C_1(r_k)']', \\
\hat{D}_{11} &= [D_{11}(r_1)' \; D_{11}(r_2)' \; \ldots \; D_{11}(r_k)']' \quad \text{and} \\
\hat{D}_{12} &= [D_{12}(r_1)' \; D_{12}(r_2)' \; \ldots \; D_{12}(r_k)']'.
\end{aligned}$$

5.3 Spatial $\mathcal{H}_\infty$ control of a piezoelectric laminate beam

This section is concerned with the application of spatial $\mathcal{H}_\infty$ control to a piezoelectric laminate beam. We consider a beam with pinned boundary conditions with a collocated piezoelectric actuator/sensor pair bonded to both its sides as shown in Figure 5.2.

Here, the voltage induced in one of the piezoelectric transducers is used as a measurement while the control voltage is applied to the other piezo-

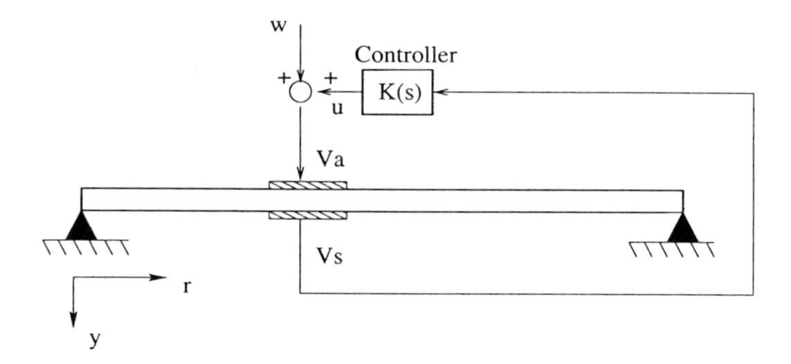

Fig. 5.2 Control of a flexible beam

electric transducer to oppose the disturbance w entering into the system through the same channel as the control signal. Dynamical properties of this system were studied in Section 2.8, where it was shown that the transfer function from the voltage applied to the actuating piezoelectric transducer, $V_a(s)$ to the transverse structural deflection $y(s,r)$ at location r is of the form

$$G(s,r) = \sum_{i=1}^{\infty} \frac{F_i \, \phi_i(r)}{s^2 + 2\zeta_i \, \omega_i \, s + \omega_i^2}. \tag{5.9}$$

Furthermore, the transfer function from the applied piezoelectric actuator voltage to the measured collocated sensor voltage is of the form

$$G_{Vs}(s) = \sum_{i=1}^{\infty} \frac{F_i \, \nu}{s^2 + 2\zeta_i \, \omega_i \, s + \omega_i^2}, \tag{5.10}$$

where ν depends on the properties of the piezoelectric laminate structure.

In practice, as explained in Chapter 2, the dynamical model of a flexible structure as described in (5.9) and (5.10) needs to be truncated to represent the system in finite dimensions. The model can be truncated so to include only the modes within the frequency bandwidth of interest. However, the truncation of the model produces additional error in the locations of the in-bandwidth zeros. This is due to the fact that the contribution of the out-of-bandwidth modes, i.e. high-frequency modes, is generally ignored in the truncation.

One way to improve the truncated model dynamics is to include feed-through terms to correct the locations of the in-bandwidth zeros. In Chapter 4 we proposed two methods to determine the feed-through term; a method based on minimizing the spatial $\mathcal{H}_\infty$ norm of the error system and another based on minimizing the spatial $\mathcal{H}_2$ norm. Since we intend to design a spatial $\mathcal{H}_\infty$ controller for the system, we choose to employ the former technique. Hence, the spatially distributed model of the system in (5.9) is approximated by $\hat{G}(s,r)$,

$$\hat{G}(s,r) = \sum_{i=1}^{N} \frac{F_i\, \phi_i(r)}{s^2 + 2\zeta_i\, \omega_i\, s + \omega_i^2} + K(r), \tag{5.11}$$

where N is the number of modes included in the model, and $K(r)$ is the feed-through term added to the truncated model to correct the locations of in-bandwidth zeros. The term $K(r)$ is then determined by minimizing the following cost function

$$J = \ll W_c(s,r)\left(G(s,r) - \hat{G}(s,r)\right) \gg_\infty^2 . \tag{5.12}$$

Here, $W_c(s,r)$ is an ideal low-pass weighting function distributed spatially over $\mathcal{R}$ with its cut-off frequency ω_c chosen to lie within the interval $\omega_c \in (\omega_N, \omega_{N+1})$.

The cost function (5.12) is minimized by setting

$$K(r) = \sum_{i=N+1}^{\infty} K_i^{opt} \phi_i(r), \tag{5.13}$$

where

$$K_i^{opt} = \frac{1}{2}\left(\frac{1}{\omega_i^2} + \frac{1}{\omega_i^2 - \omega_c^2}\right) F_i. \tag{5.14}$$

Furthermore, the infinite-dimensional model (5.10) is approximated by

$$\hat{G}_{Vs}(s) = \sum_{i=1}^{N} \frac{F_i\, \nu}{s^2 + 2\zeta_i\, \omega_i\, s + \omega_i^2} + K_{Vs} \tag{5.15}$$

using the procedure explained in Section 4.6.1. Now the beam in Figure 5.2 can be shown to have a state-space representation as,

$$\dot{x}(t) = Ax(t) + B_1 w(t) + B_2 u(t)$$
$$y(t,r) = C_1(r)x(t) + D_{11}(r)w(t) + D_{12}(r)u(t)$$
$$V_s(t) = C_2 x(t) + D_{21} w(t) + D_{22} u(t), \tag{5.16}$$

where $x \in \mathbf{R}^{2N}$ is the state, $w \in \mathbf{R}$ is the disturbance input, $u \in \mathbf{R}$ is the control input, $V_s \in \mathbf{R}$ is the measured output. Furthermore, $y \in \mathbf{R} \times [0, L]$, with L being the length of the beam, is the performance output, representing the spatial displacement at time t and at position r along the beam.

The system matrices in (5.16) can be obtained from transfer functions (5.15) and (5.11). Note that for the system shown in Figure 5.2, $D_{22} = D_{21}$ in (5.16) is the feed-through term K_{Vs} described in (5.15), while $D_{11}(r) = D_{12}(r)$ is $K(r)$ in (5.13). Moreover, $B_1 = B_2$ since disturbance is assumed to enter the system through the actuator.

Considering the above points, we may write

$$A = \begin{bmatrix} 0_{N\times N} & I_{N\times N} \\ A_{21} & A_{22} \end{bmatrix},$$

where

$$A_{21} = -\mathrm{diag}(\omega_1^2, \ldots, \omega_N^2)$$
$$A_{22} = -2\,\mathrm{diag}(\zeta_1\omega_1, \ldots, \zeta_N\omega_N)$$

and

$$B_1 = B_2 = [0 \quad \cdots \quad 0 \quad F_1 \quad \cdots \quad F_N]'$$
$$C_1(r) = [\phi_1(r) \quad \cdots \quad \phi_N(r) \quad 0 \quad \cdots \quad 0]$$
$$C_2 = \nu\,[F_1 \quad \cdots \quad F_N \quad 0 \quad \cdots \quad 0]$$
$$D_{11}(r) = D_{12}(r) = \sum_{i=N+1}^{N_{max}} K_i^{opt}\phi_i(r)$$
$$D_{21} = D_{22} = K_{Vs}. \tag{5.17}$$

Note that the feed-through terms $D_{11}(r) = D_{12}(r)$ are calculated by considering modes $N+1$ to N_{max}.

The spatial $\mathcal{H}_\infty$ control problem, depicted in Figure 5.3, is concerned with designing a controller of the form (5.2) such that the closed-loop system satisfies

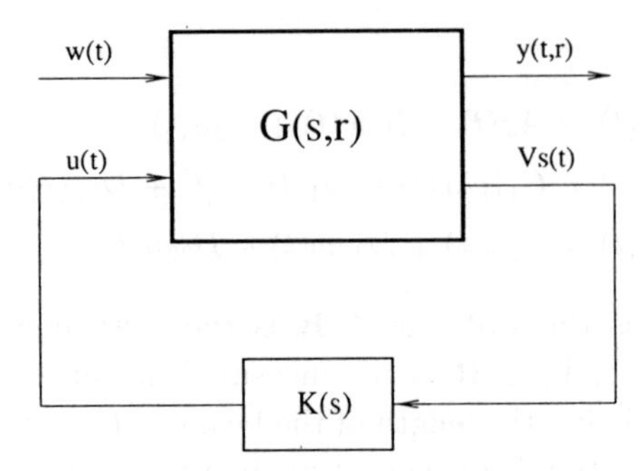

Fig. 5.3 Spatial $\mathcal{H}_\infty$ control problem

$$\inf_{K\in U} \sup_{w\in\mathcal{L}_2[0,\infty)} J_\infty < \gamma^2, \tag{5.18}$$

where U is the set of all stabilizing controllers and

$$J_\infty = \frac{\int_0^\infty \int_{\mathcal{R}} y(t,r)'y(t,r)drdt}{\int_0^\infty w(t)'w(t)dt}. \tag{5.19}$$

Following the discussion in the previous section we may write the equivalent $\mathcal{H}_\infty$ control problem as

$$\begin{aligned}
\dot{x}(t) &= Ax(t) + B_1w(t) + B_2u(t) \\
\tilde{y}(t) &= \Pi\, x(t) + \Theta w(t) + \Theta u(t) \\
V_s(t) &= C_2x(t) + D_{21}w(t) + D_{22}u(t),
\end{aligned} \tag{5.20}$$

where $D_{21} = D_{22}$ and $[\Pi \quad \Theta] = \Gamma$. Here, Γ is a matrix that satisfies

$$\Gamma'\Gamma = \int_{\mathcal{R}} \begin{bmatrix} C_1(r)' \\ D_{11}(r)' \end{bmatrix} [C_1(r) \quad D_{11}(r)]\, dr. \tag{5.21}$$

Also, $\Theta_{11} = \Theta_{12} = \Theta$ since $D_{11}(r) = D_{12}(r)$. Furthermore,

$$\Pi = \begin{bmatrix} \mathrm{diag}\,(\Phi_1, \Phi_2, \ldots, \Phi_N, 0_{1\times N}) \\ 0_{1\times 2N} \end{bmatrix} \quad \text{and}$$

$$\Theta = \begin{bmatrix} 0_{2N\times 1} \\ \left(\sum_{i=N+1}^{N_{max}} (K_i^{opt}\Phi_i)^2\right)^{\frac{1}{2}} \end{bmatrix},$$

where we have used the orthogonality condition

$$\int_{\mathcal{R}} \phi_i(r)\phi_j(r)dr = \Phi_i \delta_{ij}.$$

It can be observed that the $\mathcal{H}_\infty$ control problem associated with the system described in (5.20) is non-singular. This is due to the existence of feed-through terms from the disturbance to the measured output and from the control signal to the performance output. Had we not corrected the location of in-bandwidth zeros by adding appropriate feed-through terms to the truncated models as in (5.11) and (5.15), the resulting $\mathcal{H}_\infty$ control problem would have been singular. Furthermore, note that in the spatial $\mathcal{H}_\infty$ cost function (5.19), the spatial weight function is set to $Q(r) = 1$. This means that vibration of the entire structure is to be reduced as uniformly as possible.

Designing a $\mathcal{H}_\infty$ controller for the system (5.20) may result in a very high gain controller. This can be attributed to the fact that the term Θ in (5.20) does not represent a physical weight on the control signal. Rather, it represents the effect of truncated modes on the in-bandwidth dynamics of the system. This problem can be addressed by introducing a weight on the control signal. This can be achieved by re-writing (5.20) as,

$$\begin{aligned} \dot{x}(t) &= Ax(t) + B_1 w(t) + B_2 u(t) \\ \hat{y}(t) &= \begin{bmatrix} \Pi \\ 0 \end{bmatrix} x(t) + \begin{bmatrix} \Theta \\ 0 \end{bmatrix} w(t) + \begin{bmatrix} \Theta \\ R \end{bmatrix} u(t) \\ V_s(t) &= C_2 x(t) + D_{21} w(t) + D_{22} u(t), \end{aligned} \tag{5.22}$$

where R is a weighting matrix with compatible dimensions.

What makes this system different from (5.20) is the existence of R in the error output, $\hat{y}$. R serves as a weighting to balance the controller effort with respect to the degree of vibration reduction that can be achieved. This can be shown to be equivalent to adding a term, $\int_0^\infty u(t)'R'Ru(t)dt$, to the numerator of the cost function, J_∞ in (5.19). Setting R with smaller elements may lead to higher vibration reduction but at the expense of a higher controller gain. In practice, one has to make a compromise between the level of vibration reduction and controller gain by choosing a suitable R.

5.4 Experimental implementation of the spatial $\mathcal{H}_\infty$ controller

In this section, effectiveness of the spatial $\mathcal{H}_\infty$ control technique is demonstrated on a laboratory scale apparatus. The apparatus is the simply-supported piezoelectric laminate beam described in the previous section whose physical properties are described in Section 4.8.

Our goal is to control only the first six vibration modes of the beam using a SISO controller. Since we wish to implement a controller with the smallest possible number of states, the model is truncated to include only the first six bending modes of the structure, i.e. $N = 6$. The $\mathcal{H}_\infty$ control design procedure will then produce a 12th order controller.

The effect of out-of-bandwidth modes is incorporated into the model by adding appropriate feed-through terms, as explained in the previous section. Based on the experimental frequency-response data from actuator voltage to sensor voltage, the feed-through term in (5.15), $D_{21} = D_{22} = K_{Vs}$, is found to be 0.033. For the spatially distributed model $G(s, r)$, the feed-through term is calculated by considering modes $N+1 = 7$ to $N_{max} = 200$.

In our design the scalar weighting factor is set to $R = 4.78 \times 10^{-7}$. This choice results in a controller with sufficient damping and robustness properties. Mat-lab μ-Analysis and Synthesis Toolbox was used to calculate a spatial $\mathcal{H}_\infty$ controller based on the system in (5.22). The optimum upper bound on the cost function was found to be $\gamma_{opt}^2 = 9.6 \times 10^{-6}$.

The experimental setup is depicted in Figure 5.4. The controller was implemented using a dSpace DS1103 rapid prototyping system together with the Matlab and Simulink software. The sampling frequency was set at $T_s = 20$ KHz, while the cut-off frequencies of the two low-pass filters were set at 3 KHz each. A high voltage amplifier, capable of driving highly capacitive loads, was used to supply voltage for the actuating piezoelectric patch. An HP89410A Dynamic Signal Analyzer and a Polytec PSV-300 Laser Doppler Scanning Vibrometer were used to collect frequency domain data from the piezoelectric laminate beam. Important parameters of the beam, such as resonance frequencies and damping ratios, were obtained from the experiment and were used to correct our models.

The frequency response of the controller is shown in Figure 5.5. It can be observed that the controller has a highly resonant nature. This is expected and can be attributed to the highly resonant nature of the structure. That is, the controller attempts to apply a high gain at each resonance frequency. Figures 5.6 and 5.7 compare frequency responses of the open-loop and closed-loop systems (actuator voltage to sensor voltage).

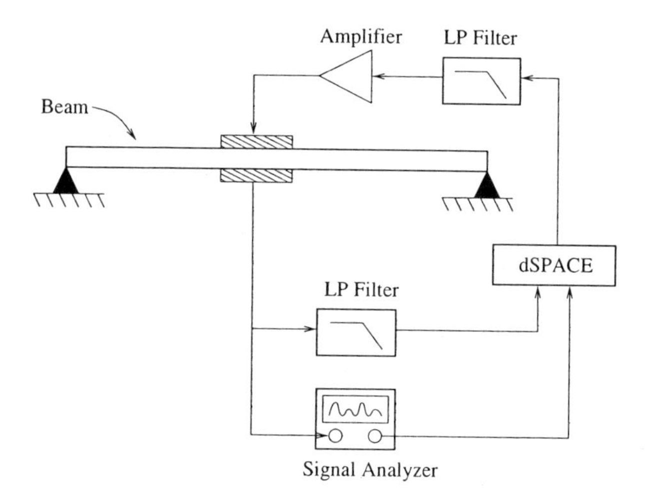

Fig. 5.4 Experimental setup

Both simulation and experimental results are plotted. It can be observed that the performance of the controller applied to the real system is as expected. The resonant responses of the first six modes of the system have been considerably reduced once the controller was introduced.

Figures 5.8 and 5.9 show the simulated spatial frequency responses of the uncontrolled and controlled beam respectively. Here, r is measured from one end of the beam, which is closer to the patches. We also obtained the spatial frequency responses from the experiments. A Polytec PSV-300 Laser Scanning Vibrometer was used to obtain the frequency response of the beam at a number of points on the surface of the beam. The spatial frequency responses of the uncontrolled and controlled beam are shown in Figures 5.10 and 5.11. It can be observed that the resonant responses of modes $1-6$ have been reduced over the entire beam as a result of the controller action. Furthermore, it can be observed that the experimental results are very close to the simulations. The contour plots of the experimental spatial frequency responses are shown in Figure 5.12. The contour plots are at levels between -100 dB and -190 dB with an interval of 10 dB.

The results show that the resonant responses of modes $1-6$ have been reduced by approximately $27, 30, 19.5, 19.5, 15.5$ and 8 dB respectively over the entire beam. Thus, our spatial $\mathcal{H}_\infty$ controller minimizes resonant responses of selected vibration modes over the entire structure, which is desirable for vibration suppression purposes.

To demonstrate the controller effect on the spatial $\mathcal{H}_\infty$ norm of the system, we have plotted the point-wise $\mathcal{H}_\infty$ norm of the controlled and uncontrolled beam as a function of r in Figure 5.13. The figures show that the experimental results are very similar to the simulations. Furthermore, they clearly demonstrate the effect of our spatial $\mathcal{H}_\infty$ controller in reducing the vibration of the beam. It is obvious that the $\mathcal{H}_\infty$ norm over the entire beam has been reduced by the action of the controller in a uniform manner. The highest $\mathcal{H}_\infty$ norm of the uncontrolled beam has been reduced by approximately 97%, from 3.6×10^{-5} to 1.1×10^{-6}.

The effectiveness of the controller in minimizing the vibrations of the beam in time domain can be seen in Figure 5.14. A pulse-shaped disturbance signal was applied through the piezoelectric actuator. The disturbance had a duration of 15 seconds and and amplitude of 100 volts. The velocity response at a point 80 mm away from one end of the beam was observed using the PSV Laser Vibrometer. The velocity response was filtered by a band-pass filter from 10 Hz to 750 Hz. The settling time of the velocity response was reduced considerably.

To show the advantage of the spatial $\mathcal{H}_\infty$ control over the point-wise $\mathcal{H}_\infty$ control, we performed the following experiment. A point-wise $\mathcal{H}_\infty$ controller was designed to minimize the deflection at the middle of the beam, i.e. $r = 0.3$ m. The controller had a gain margin of 14.3 dB and a phase margin of 77.9^o. It was implemented on the beam using the setup in Figure 5.4. In Figure 5.15, we have plotted $\mathcal{H}_\infty$ norm of the controlled and uncontrolled beam as a function of r.

Figure 5.15 demonstrates the effectiveness of the point-wise control in local reduction of the $\mathcal{H}_\infty$ norm at and around $r = 0.3$ m. This is not surprising as the only purpose of the controller is to minimize vibration at $r = 0.3$ m. In fact, the point-wise controller only suppresses the odd numbered modes since $r = 0.3$ m is a node for even numbered modes of the beam. It is evident from Figures 5.13 and 5.15 that the spatial $\mathcal{H}_\infty$ controller has an advantage over the point-wise $\mathcal{H}_\infty$ controller as it minimizes the vibration throughout the entire structure.

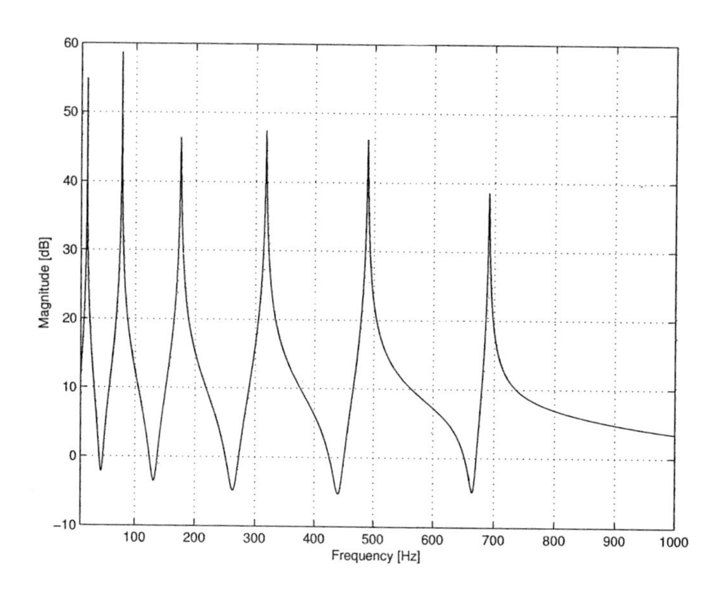

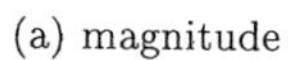

(a) magnitude

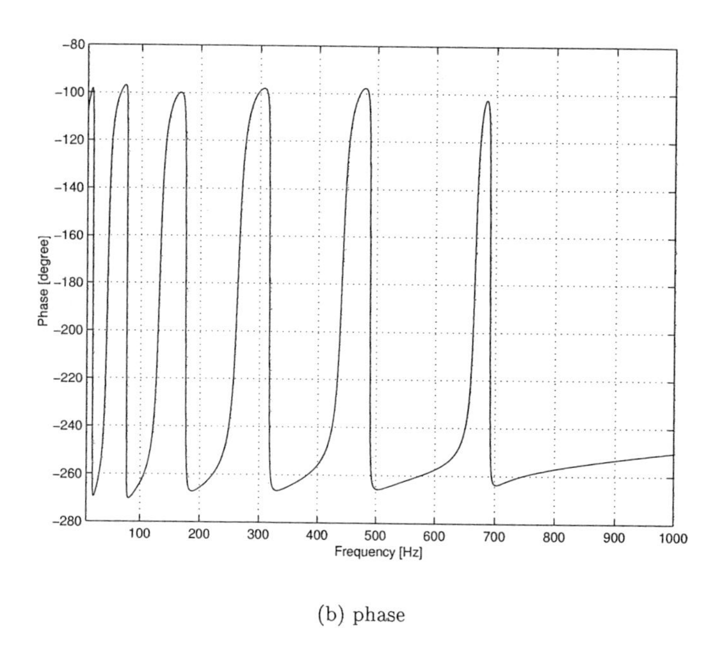

(b) phase

Fig. 5.5 Frequency response of the controller (input voltage to output voltage [V/V])

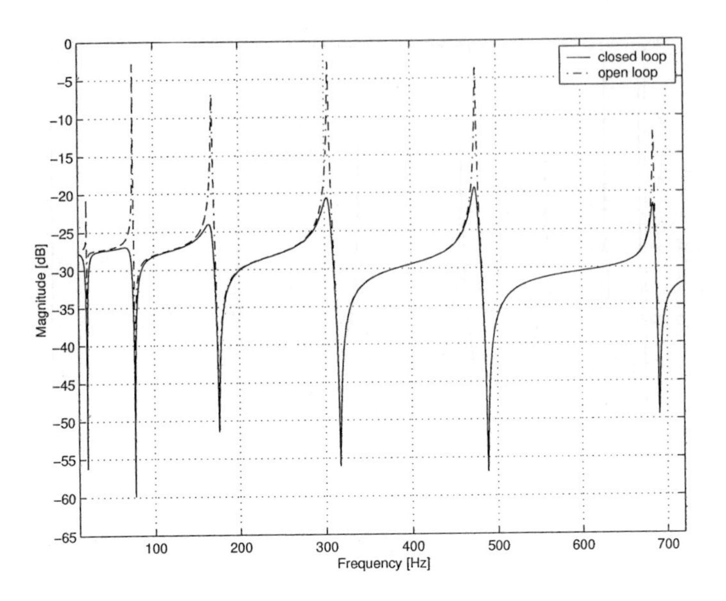

(a) simulation

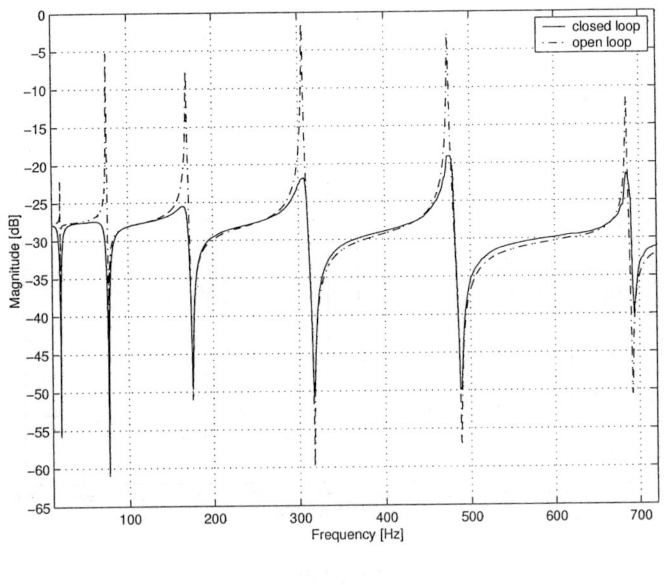

(b) experiment

Fig. 5.6 Simulated and experimental frequency responses (actuator voltage to sensor voltage [V/V]) - Magnitude

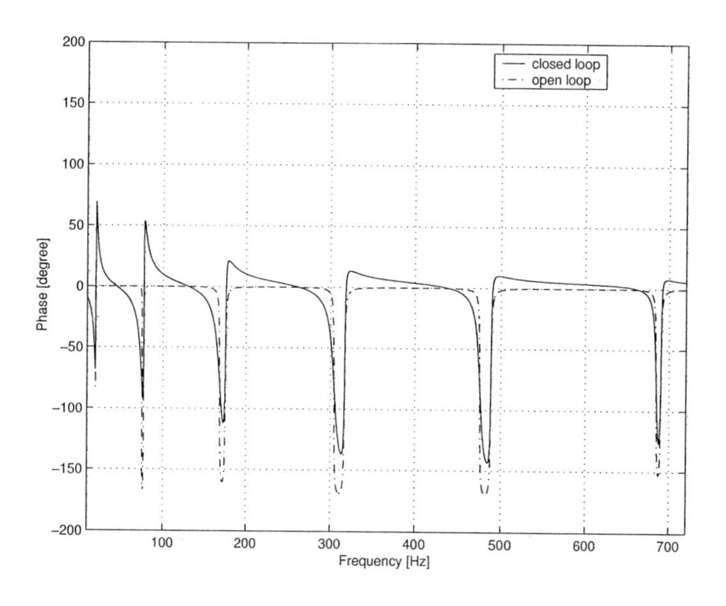

(a) simulation

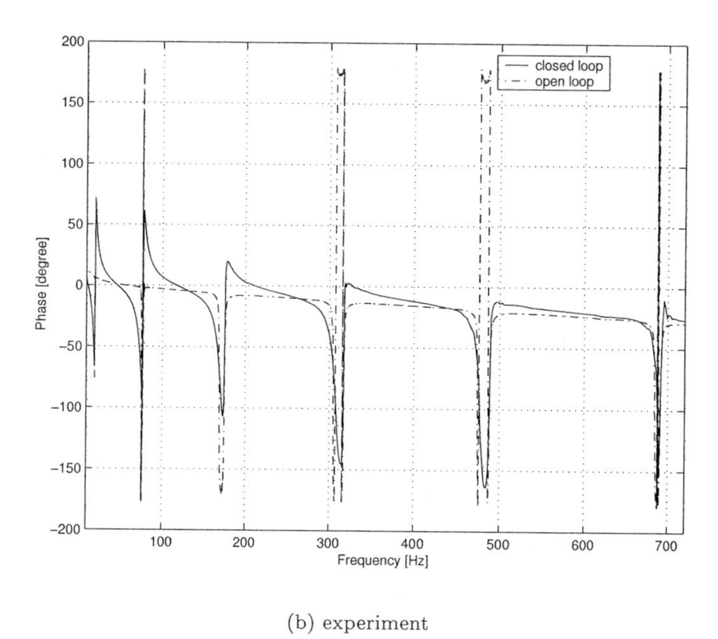

(b) experiment

Fig. 5.7 Simulated and experimental frequency responses (actuator voltage to sensor voltage [V/V]) - Phase

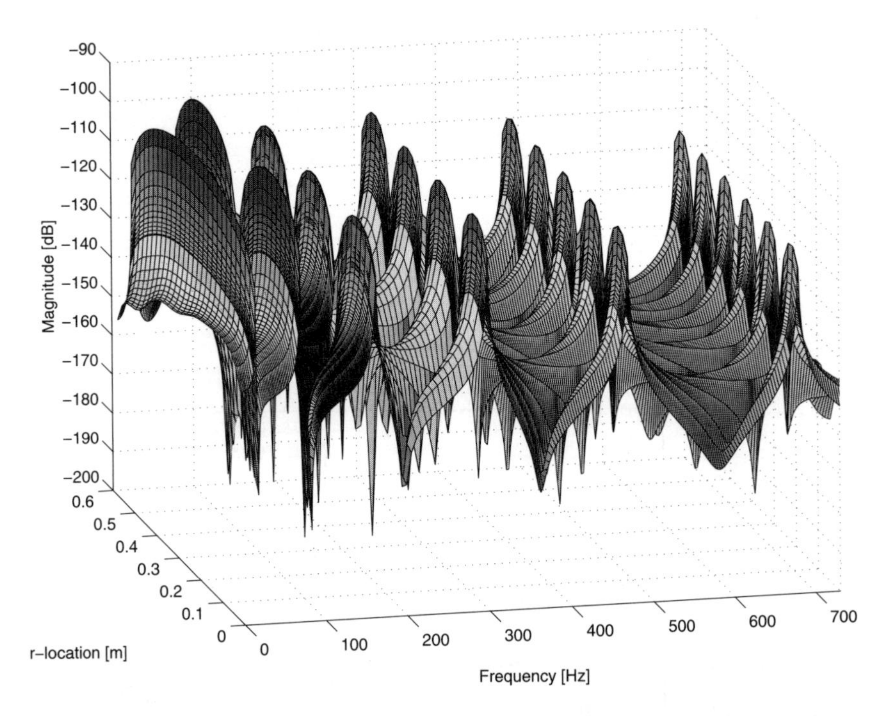

Fig. 5.8 Simulated spatial frequency response: (open loop) actuator voltage to beam deflection [m/V]

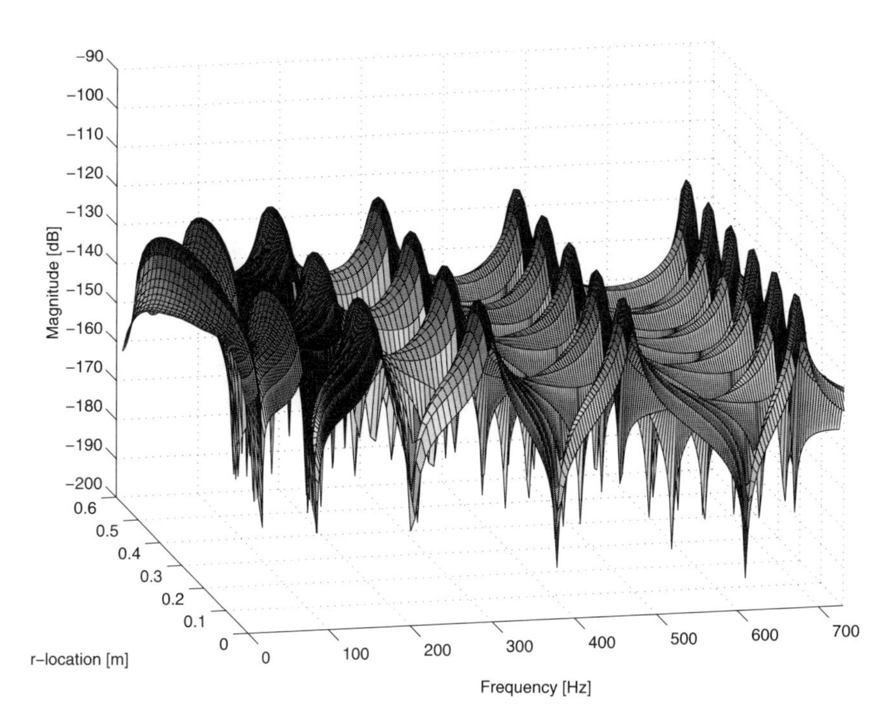

Fig. 5.9 Simulated spatial frequency response: actuator voltage - beam deflection (closed loop) [m/V]

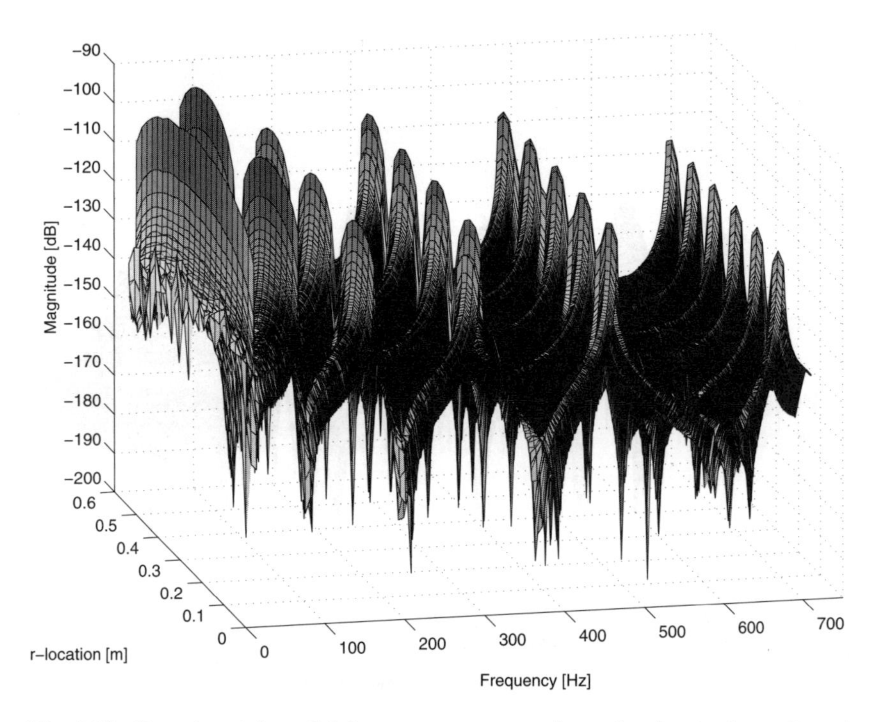

Fig. 5.10 Experimental spatial frequency response: (open loop) actuator voltage to beam deflection [m/V]

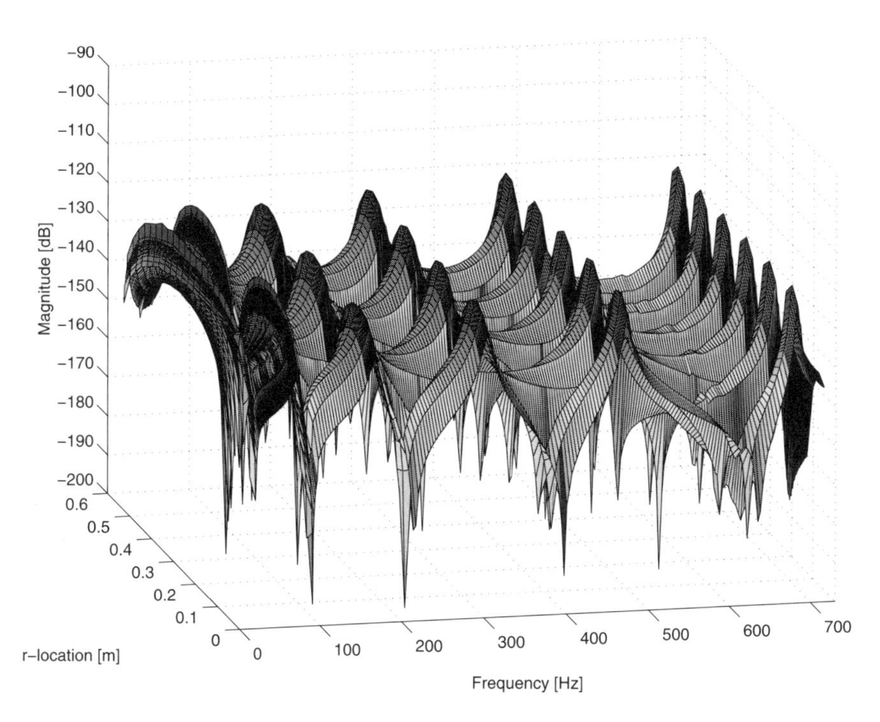

Fig. 5.11 Experimental spatial frequency response: actuator voltage - beam deflection (closed loop) [m/V]

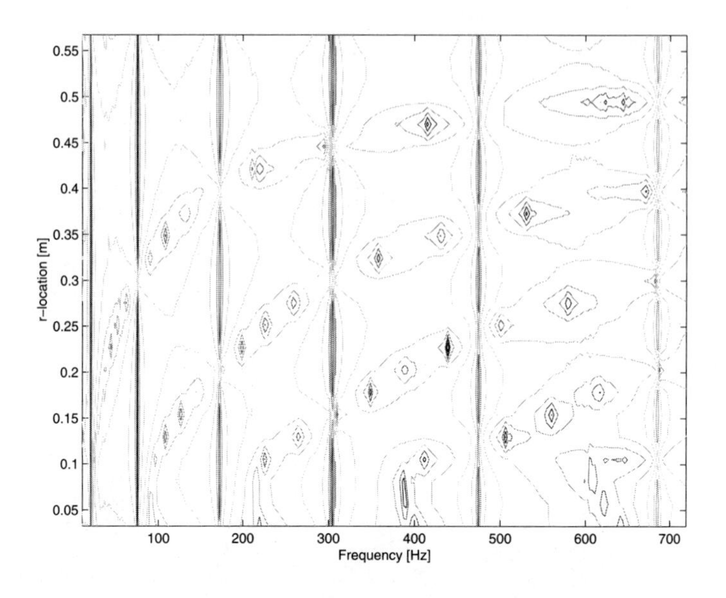

(a) open loop

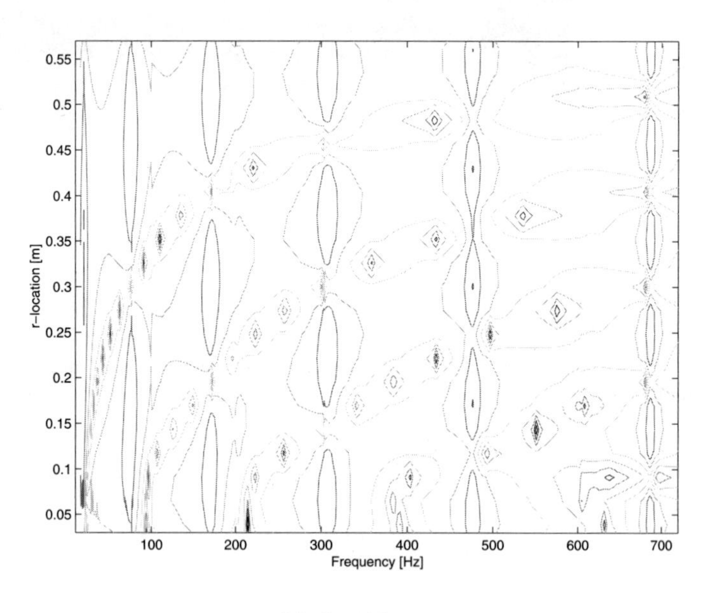

(b) closed loop

Fig. 5.12 Contour plots from experiments

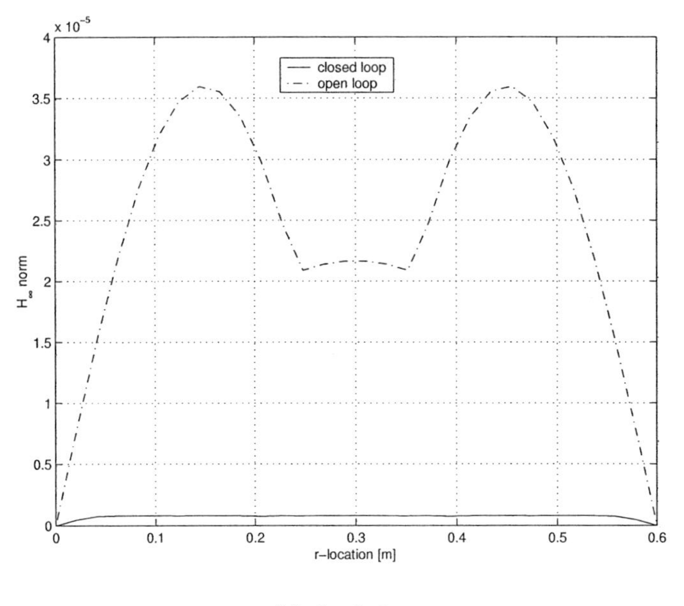

(a) simulation

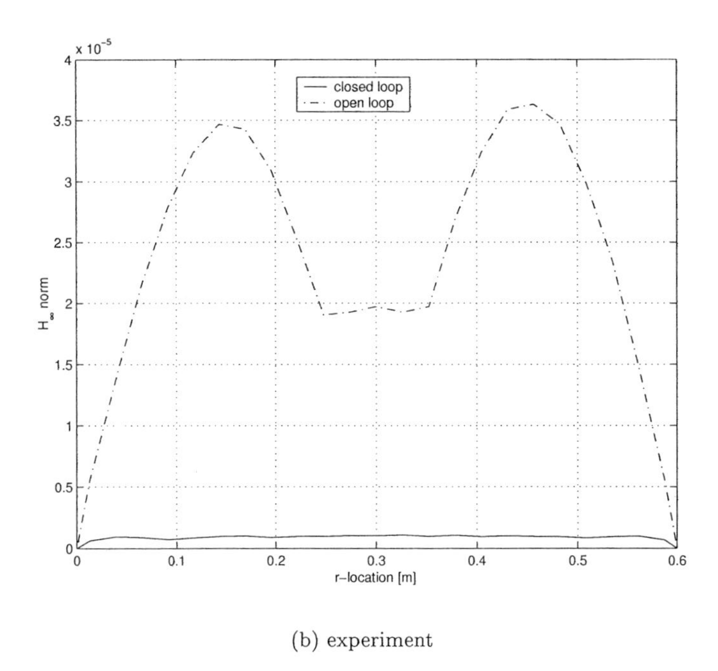

(b) experiment

Fig. 5.13 Simulated and experimental $\mathcal{H}_\infty$ norm plot - spatial control

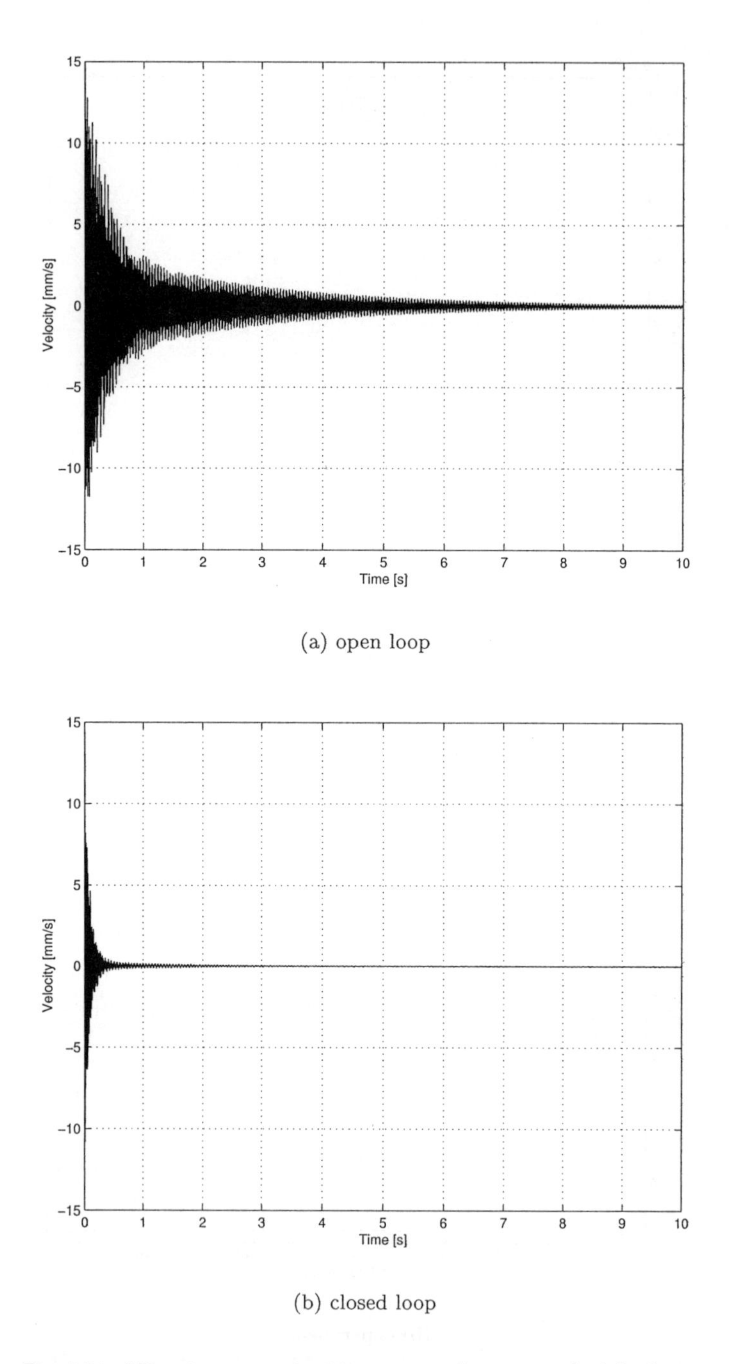

Fig. 5.14 Vibration at a point 80 mm away from one end of the beam

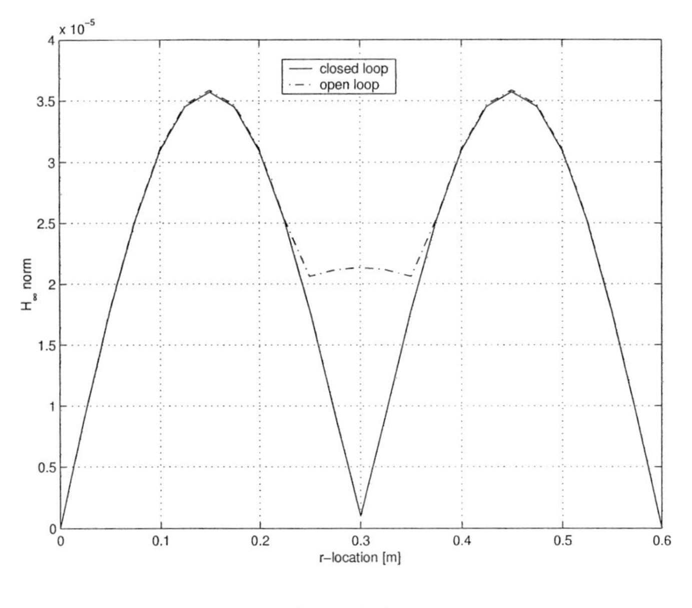

(a) simulation

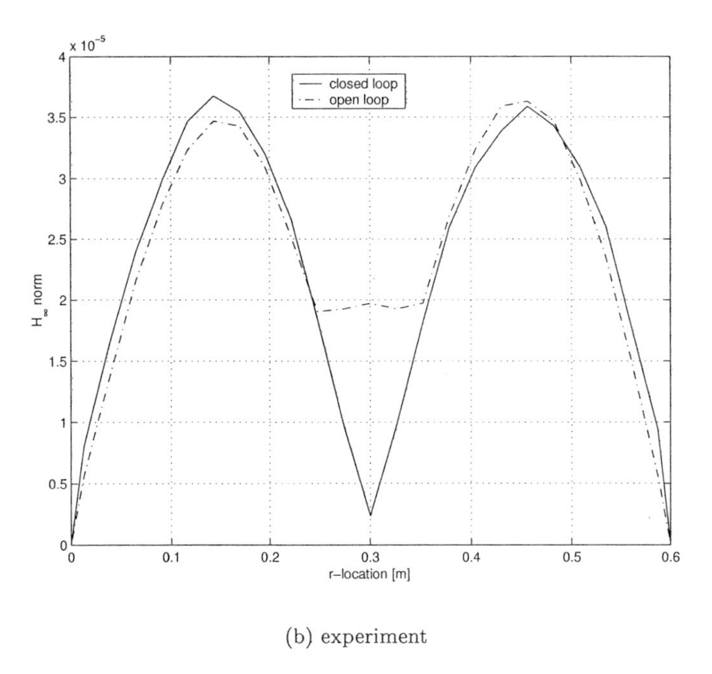

(b) experiment

Fig. 5.15 Simulated and experimental $\mathcal{H}_\infty$ norm plot - point-wise control

5.5 The effect of pre-filtering on performance of the spatial $\mathcal{H}_\infty$ controller

Spatial $\mathcal{H}_\infty$ controllers inherit many of the properties of the conventional $\mathcal{H}_\infty$ controllers. In this section we consider one such property that could be useful in further suppressing the in-bandwidth vibration modes at the expense of possibly increasing out-of-bandwidth vibrations.

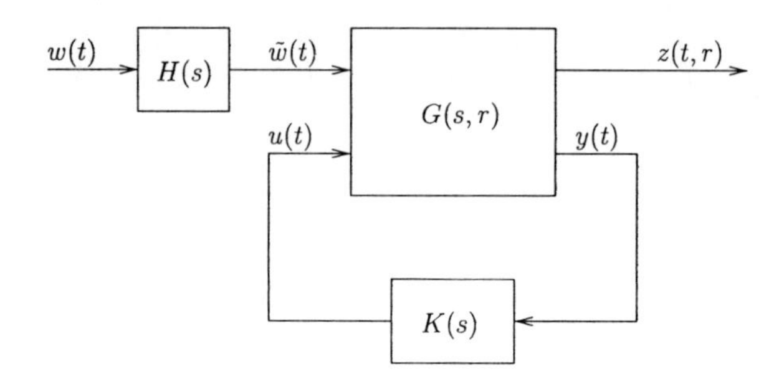

Fig. 5.16 Block diagram representation of a spatial $\mathcal{H}_\infty$ control problem with pre-filtered disturbance

Let us consider the system in Figure 5.16 and let $\mathcal{S}_K$ denote the linear operator mapping w to z. That is, $\mathcal{S}_K$ corresponds to a transfer function matrix $S_K(s,r)$. Also let $G_K(s,r)$ represent the closed-loop transfer function matrix from $\tilde{w}(t)$ to $z(t,r)$ and assume that G_K has a state-space realization

$$\begin{aligned} \dot{x}(t) &= Ax(t) + B\tilde{w}(t) \\ z(t,r) &= C(r)x(t). \end{aligned}$$

Furthermore, allow $H(s)$ to be a given weighting function with state-space realization

$$\begin{aligned} \dot{x}_H(t) &= A_H x_H(t) + B_H w(t) \\ \tilde{w}(t) &= C_H x_H(t). \end{aligned}$$

The following theorem states the effect of including a pre-filter $H(s)$ in the controller design on the closed-loop spatial frequency response of the system.

Theorem 5.1 *Consider the system of Figure 5.16. The following statements are equivalent:*

(1) $\ll \mathcal{S}_K \gg < \gamma$.
(2) $\ll G_K(j\omega, r) \gg < \frac{\gamma}{\|H(j\omega)\|} \quad \forall \omega \in \mathbf{R}$.

Here, $\ll G_K(j\omega, r) \gg = \left(\int_{\mathcal{R}} G_K(j\omega, r)^* G_K(j\omega, r) dr \right)^{\frac{1}{2}}$.

Proof

Consider Figure 5.16, we can write

$$S_K(s, r) = G_K(s, r) H(s).$$

Therefore, $S_K(s, r)$ will have the following state-space realization

$$\begin{aligned} \dot{\hat{x}}(t) &= \hat{A}\hat{x}(t) + \hat{B}w(t) \\ z(t, r) &= \hat{C}(r)\hat{x}(t), \end{aligned}$$

where

$$\begin{aligned} \hat{A} &= \begin{bmatrix} A & BC_H \\ 0 & A_H \end{bmatrix}; \\ \hat{B} &= \begin{bmatrix} 0 \\ B_H \end{bmatrix}; \\ \hat{C}(r) &= \begin{bmatrix} C(r) & 0 \end{bmatrix}; \quad \text{and} \\ \hat{x} &= \begin{bmatrix} x \\ x_H \end{bmatrix}. \end{aligned}$$

Corresponding to this system is the finite-dimensional system $\tilde{S}_K(s)$ with an equivalent $\mathcal{H}_\infty$ norm, i.e.

$$\ll S_K(s, r) \gg_\infty = \|\tilde{S}_K(s)\|_\infty$$

which has the following state-space realization:

$$\begin{aligned} \dot{\hat{x}}(t) &= \hat{A}\hat{x}(t) + \hat{B}w(t) \\ \tilde{z}(t) &= \Pi\hat{x}(t), \end{aligned}$$

where

$$\Pi = \left[\left(\int_{\mathcal{R}} C(r)' C(r) dr \right)^{\frac{1}{2}} \; 0 \right].$$

It is straightforward to verify that

$$\tilde{S}_K(s) = \tilde{G}(s) \times H(s),$$

where $\tilde{G}_K(s)$ is the finite-dimensional counterpart of $G_K(s,r)$ with the following state-space realization:

$$\begin{aligned}\dot{\tilde{x}}(t) &= A\tilde{x}(t) + Bw(t)\\ \tilde{z}(t) &= \Pi\tilde{x}(t).\end{aligned}$$

Now, if we assume that $\ll \mathcal{S}_K \gg < \gamma$ we have

$$\begin{gathered}\ll \mathcal{S}_K \gg < \gamma\\ \Updownarrow\\ \|\tilde{S}_K(s)\|_\infty < \gamma\\ \Updownarrow\\ \|\tilde{G}_K(s)H(s)\|_\infty < \gamma\\ \Updownarrow\\ \|\tilde{G}_K(j\omega)\| < \tfrac{\gamma}{\|H(j\omega)\|} \quad \forall\omega \in \mathbf{R}\\ \Updownarrow\\ \ll G_K(j\omega, r) \gg < \tfrac{\gamma}{\|H(j\omega)\|} \quad \forall\omega \in \mathbf{R}\end{gathered}$$

which proves the statement of the theorem. □

5.6 The spatial $\mathcal{H}_2$ control problem

This section is concerned with the problem of spatial $\mathcal{H}_2$ control of spatially distributed systems of the form

$$\begin{aligned}\dot{x}(t) &= Ax(t) + B_1w(t) + B_2u(t)\\ z(t,r) &= C_1(r)x(t) + D_{12}(r)u(t)\\ y(t) &= C_2x(t) + D_{21}w(t) + D_{22}u(t),\end{aligned} \tag{5.23}$$

where $r \in \mathcal{R}$ is the spatial coordinate, $x \in \mathbf{R}^n$ is the state of the system, $w \in \mathbf{R}^p$ is the disturbance input, $u \in \mathbf{R}^m$ is the control input, $z \in \mathbf{R}^q \times \mathcal{R}$ is the performance output and $y \in \mathbf{R}^\ell$ is the measured output.

The spatial $\mathcal{H}_2$ control problem is to design a full-order controller

$$\begin{aligned}\dot{x}_k(t) &= A_kx_k(t) + B_ky(t)\\ u(t) &= C_kx_k(t) + D_ky(t)\end{aligned} \tag{5.24}$$

such that the weighted spatial $\mathcal{H}_2$ norm of the closed-loop system

$$\ll T_{zw}(s,r) \gg_{2,Q}^2 = \frac{1}{2\pi}\int_{-\infty}^{\infty}\int_{\mathcal{R}} \mathrm{tr}\{T_{zw}(j\omega,r)^* Q(r) T_{zw}(j\omega,r)\} dr\, d\omega \tag{5.25}$$

is minimized.

Here, $Q(r)$ is a spatial weighting function and T_{zw} is the closed-loop transfer function from w to z. The purpose of $Q(r)$ is to emphasize the region over which the spatial $\mathcal{H}_2$ norm of the system is to be minimized more heavily.

It can be verified that the above problem is equivalent to a standard $\mathcal{H}_2$ control problem for the system,

$$\begin{aligned} \dot{x}(t) &= Ax(t) + B_1 w(t) + B_2 u(t) \\ \tilde{z}(t) &= \Pi x(t) + \Theta_{12} u(t) \\ y(t) &= C_2 x(t) + D_{21} w(t) + D_{22} u(t) \end{aligned} \tag{5.26}$$

and the cost function

$$||\tilde{T}_{zw}(s)||_2^2 = \frac{1}{2\pi}\int_{-\infty}^{\infty} \mathrm{tr}\{\tilde{T}_{zw}(j\omega)^* T_{zw}(j\omega)\}\, d\omega, \tag{5.27}$$

where

$$\left[\Pi\ \Theta_{12}\right] = \Gamma$$

and Γ is a matrix that satisfies

$$\Gamma'\Gamma = \int_{\mathcal{R}} \begin{bmatrix} C(r)' \\ D_{12}(r)' \end{bmatrix} Q(r) \left[C(r)\ D_{12}(r)\right] dr. \tag{5.28}$$

Hence, the spatial $\mathcal{H}_2$ control problem as defined by (5.23) and (5.25) is equivalent to a standard $\mathcal{H}_2$ control problem defined by (5.26), (5.27) and (5.28). Therefore, the spatial $\mathcal{H}_2$ control problem can be solved using standard computational software available for standard $\mathcal{H}_2$ control.

If the purpose of the controller is to minimize the $\mathcal{H}_2$ norm at a number of specific points inside $\mathcal{R}$, then $Q(r)$ should be constructed from a number of Dirac delta functions each centered at a specific point. That is,

$$Q(r) = \sum_{i=1}^{k} \delta(r - r_i). \tag{5.29}$$

Then, it can be verified that the spatial $\mathcal{H}_2$ control problem (5.23), (5.25) reduces to the problem defined by the system

$$\begin{aligned}\dot{x}(t) &= Ax(t) + B_1 w(t) + B_2 u(t)\\ \hat{z}(t) &= \hat{C}_1 x(t) + \hat{D}_{12} u(t)\\ y(t) &= C_2 x(t) + D_{21} w(t) + D_{22} u(t),\end{aligned} \tag{5.30}$$

and the performance objective

$$\|T_{\hat{z}w}\|_2^2 = \frac{2}{\pi}\int_{-\infty}^{\infty} \operatorname{tr}\{T_{\hat{z}w}(j\omega)^* T_{\hat{z}w}(j\omega)\} d\omega, \tag{5.31}$$

where matrices $\hat{C}_1$ and $\hat{D}_{12}$ in (5.30) are given below:

$$\begin{aligned}\hat{C}_1 &= [C_1(r_1)'\ C_1(r_2)'\ \dots\ C_1(r_k)']'\\ \hat{D}_{12} &= [D_{12}(r_1)'\ D_{12}(r_2)'\ \dots\ D_{12}(r_k)']'.\end{aligned}$$

Designing a $\mathcal{H}_2$ controller for the system (5.26) may result in a very high gain controller, which may not have the required robustness properties. This problem can be avoided by introducing a weight factor, R, on the control signal. This can be achieved by re-writing (5.26) as

$$\begin{aligned}\dot{x}(t) &= Ax(t) + B_1 w(t) + B_2 u(t)\\ \tilde{z}(t) &= \begin{bmatrix}\Pi\\ 0\end{bmatrix} x(t) + \begin{bmatrix}\Theta_{12}\\ R\end{bmatrix} u(t)\\ y(t) &= C_2 x(t) + D_{21} w(t) + D_{22} u(t).\end{aligned} \tag{5.32}$$

In practice, one has to make a compromise between the level of vibration reduction and controller gain by choosing a suitable R.

5.7 Spatial $\mathcal{H}_2$ control of a piezoelectric laminate beam

In this section we address the problem of the spatial $\mathcal{H}_2$ control of the piezoelectric laminate beam discussed in Section 5.3.

A spatially-distributed model of the system is given in (5.16). There is a difficulty in using this model for the purpose of designing a spatial $\mathcal{H}_2$ controller since the feed-through term $D_{12}(r)$ is non-zero. To alleviate this problem, we may replace the feed-through term with a second order out-of-bandwidth term as suggested in [11]. We have to ensure that the second order term is added such that the zero-frequency content of the resulting system is close to that of (5.16). The resonance frequency of the second-order system, ω_c, is set above the bandwidth of interest. Also, a relatively high damping ratio, ζ_c, is used, so that the second-order system behaves like a low-pass filter.

Thus the modified system can be expressed as

$$\begin{aligned}\dot{\tilde{x}}(t) &= \tilde{A}\tilde{x}(t) + \tilde{B}_1 w(t) + \tilde{B}_2 u(t)\\ y(t,r) &= \tilde{C}_1(r)\tilde{x}(t)\\ V_s(t) &= \tilde{C}_2\tilde{x}(t) + D_{21}w(t) + D_{22}u(t),\end{aligned} \tag{5.33}$$

where $\tilde{x}$ consists of the original states of the plant, x, and the states of the extra second-order term. Also,

$$\tilde{A} = \begin{bmatrix} 0_{(N+1)\times(N+1)} & I_{(N+1)\times(N+1)} \\ \tilde{A}_{1_{(N+1)\times(N+1)}} & \tilde{A}_{2_{(N+1)\times(N+1)}} \end{bmatrix},$$

where

$$\begin{aligned}\tilde{A}_1 &= -\mathrm{diag}(\omega_1^2, \ldots, \omega_N^2, \omega_c^2)\\ \tilde{A}_2 &= -2\,\mathrm{diag}(\zeta_1\omega_1, \ldots, \zeta_N\omega_N, \zeta_c\omega_c)\end{aligned}$$

and

$$\begin{aligned}\tilde{B}_1 &= \tilde{B}_2 = [0, \cdots, 0, 0, F_1, \cdots, F_N, 1]'\\ \tilde{C}_1(r) &= \left[\phi_1(r), \cdots, \phi_N(r), \omega_c^2 \sum_{i=N+1}^{N_{max}} K_i^{opt}\phi_i(r), 0, \cdots, 0, 0\right]\\ \tilde{C}_2 &= \nu\,[F_1, \cdots, F_N, 0, 0, \cdots, 0, 0]\\ D_{21} &= D_{22} = K_{Vs}.\end{aligned} \tag{5.34}$$

Notice that $K_{N+1}^{opt}, \ldots, K_{N_{max}}^{opt}$ are determined using the procedure described in Section 4.3. That is, the spatially distributed feed-through term minimizes the spatial $\mathcal{H}_2$ norm of the error system.

Then, the spatial $\mathcal{H}_2$ control problem is to design a controller of the form (5.24) such that the closed-loop system minimizes the weighted spatial $\mathcal{H}_2$ norm of the closed-loop system:

$$\begin{aligned}&\ll T_{yw}(s,r) \gg_{2,Q}^2 =\\ &\frac{1}{2\pi}\int_{-\infty}^{\infty}\int_{\mathcal{R}} \mathrm{tr}\{T_{yw}(j\omega,r)^* Q(r) T_{yw}(j\omega,r)\}dr\,d\omega.\end{aligned} \tag{5.35}$$

In this particular application the spatial weighting function is chosen to be $Q(r) = 1$. In other words, the entire beam is weighted equally. This reflects our interest to minimize vibration of the entire structure in a uniform manner.

This problem can then be shown to be equivalent to a standard $\mathcal{H}_2$ control problem for the following system:

$$\begin{aligned}\dot{\tilde{x}}(t) &= \tilde{A}\tilde{x}(t) + \tilde{B}_1 w(t) + \tilde{B}_2 u(t)\\ \tilde{y}(t) &= \Gamma\,\tilde{x}(t)\\ V_s(t) &= \tilde{C}_2\tilde{x}(t) + D_{21}w(t) + D_{22}u(t),\end{aligned} \tag{5.36}$$

where $D_{21} = D_{22}$ and Γ is any matrix that satisfies

$$\Gamma'\Gamma = \int_{\mathcal{R}} \tilde{C}_1(r)'\,\tilde{C}_1(r)\,dr. \tag{5.37}$$

We can obtain Γ using the orthonormality property of the eigenfunction, ϕ_i,

$$\Gamma = \begin{bmatrix}\tilde{\Gamma}_{(N+1)\times(N+1)} & 0_{(N+1)\times(N+1)}\\ 0_{(N+1)\times(N+1)} & 0_{(N+1)\times(N+1)}\end{bmatrix},$$

where

$$\tilde{\Gamma} = \operatorname{diag}\left(1,\ldots,1,\omega_c^2\left(\Sigma_{i=N+1}^{N_{max}}(K_i^{opt})^2\right)^{\frac{1}{2}}\right).$$

Finally, to ensure that the $\mathcal{H}_2$ controller design for the system (5.36) will not result in a very high gain controller, we may introduce a weight factor, R, on the control signal. This can be achieved by re-writing (5.36) as,

$$\begin{aligned}\dot{\tilde{x}}(t) &= \tilde{A}\tilde{x}(t) + \tilde{B}_1 w(t) + \tilde{B}_2 u(t)\\ \hat{y}(t) &= \begin{bmatrix}\Gamma\\ 0\end{bmatrix}\tilde{x}(t) + \begin{bmatrix}0\\ R\end{bmatrix}u(t)\\ V_s(t) &= \tilde{C}_2\tilde{x}(t) + D_{21}w(t) + D_{22}u(t).\end{aligned} \tag{5.38}$$

5.8 Experimental implementation of spatial $\mathcal{H}_2$ control

In this section we report experimental implementation of a spatial $\mathcal{H}_2$ controller on our piezoelectric laminate beam structure. The experimental setup is depicted in Figure 5.4. The controller was implemented using a dSpace DS1103 rapid prototyping system together with the Matlab and Simulink software. The sampling frequency was set at 20 KHz, while the cut-off frequencies of the two low-pass filters were set at 3 KHz. The equipment used are similar to the ones mentioned in Section 5.4.

The frequency response of the controller is shown in Figure 5.17. The controller has a resonant nature which is largely due to the fact that the underlying system consists of a number of lightly damped modes. That is, the controller attempts to apply a high gain at each resonance frequency to minimize the resonant response in a spatial $\mathcal{H}_2$ sense. Figures 5.18 and 5.19 compare frequency responses of the open-loop and closed-loop systems (actuator voltage to sensor voltage). Both simulation and experimental results are plotted. The performance of the controller applied to the real system is as expected from the simulation.

Figures 5.20 and 5.21 show the simulated spatial frequency responses of the uncontrolled and controlled beam respectively. Here, r is measured from one end of the beam, which is closer to the piezoelectric patches. Spatial frequency responses were obtained experimentally at a number of points on the surface of the beam using a Polytec PSV-300 Laser Scanning Vibrometer. The 3-D frequency responses of the uncontrolled and controlled beam are shown in Figures 5.22 and 5.23. It can be observed that the resonant responses of the first six modes of the beam have been reduced over the entire beam due to the controller action. Comparing these results with Figures 5.20 and 5.21, we notice that the experimental results are very similar to the simulated plots. The experimental contour plots of the spatial frequency responses are shown in Figure 5.24. The plots show the contour at levels between -100 dB and -190 dB with an interval of 10 dB. The resonant responses of modes $1-6$ have been reduced by approximately $25.5, 28.5, 18, 18, 14$ and 7 dB respectively over the entire beam. Similar to the spatial $\mathcal{H}_\infty$ control, the amount of vibration reduction is greater for low-frequency modes than for high-frequency modes. This is beneficial since low-frequency modes are often the significant contributors to vibrations of flexible structures.

To show the advantage of the spatial $\mathcal{H}_2$ control over the point-wise $\mathcal{H}_2$ control, we performed the following simulation. Based on our spatial $\mathcal{H}_2$ controller, we have plotted the $\mathcal{H}_2$ norm of the controlled and uncontrolled beam as a function of r in Figure 5.25 (a). Next, a point-wise $\mathcal{H}_2$ controller was designed to minimize the deflection at the middle of the beam, i.e. $r = 0.3$ m. The controller had a gain margin of 12.7 dB and a phase margin of 87.6^o. Based on this point-wise controller, we have also plotted the $\mathcal{H}_2$ norm of the controlled and uncontrolled beam as a function of r in Figure 5.25 (b).

Figure 5.25 (a) clearly demonstrates the effect of our spatial $\mathcal{H}_2$ controller in reducing vibration of the beam. It is obvious that the $\mathcal{H}_2$ norm of the entire beam has been reduced by the action of the controller in a more uniform manner. The highest $\mathcal{H}_2$ norm of the uncontrolled beam has been reduced by approximately 69.5%, from 2.95×10^{-5} to 9.0×10^{-6}.

Meanwhile, Figure 5.25 (b) shows the effectiveness of the point-wise control in local reduction of the $\mathcal{H}_2$ norm, especially at and around $r = 0.3$ m. This is expected since the purpose of this controller is to minimize vibration at $r = 0.3$ m. Similar to the case of the point-wise $\mathcal{H}_\infty$ controller, the point-wise $\mathcal{H}_2$ controller only suppresses the odd numbered modes since $r = 0.3$ m is a node for even numbered modes. Comparing Figures 5.25 (a) and (b), it can be concluded that the spatial $\mathcal{H}_2$ controller has an advantage over the point-wise $\mathcal{H}_2$ controller in minimizing the structural vibration spatially.

Figure 5.26 demonstrates the effectiveness of the controller in minimizing beam vibration in the time domain. A step disturbance signal with an amplitude of 100 volts and a duration of 15 seconds was applied through the piezoelectric actuator. The velocity response at the middle of the beam was observed and recorded using the PSV Laser Vibrometer. The velocity response was filtered by a low-pass filter with a cut-off frequency of 750 Hz. It can be seen that the settling time of the velocity response has been substantially reduced as a result of the controller action.

5.9 Conclusions

The problem of spatial control was introduced in this chapter. We argued that given the spatially distributed nature of flexible structures, designing a controller to minimize vibration at a specific point will not necessarily improve the vibration profile of the structure elsewhere. This point was demonstrated via simulation and experimentation. Two control design methodologies based on minimizing the spatial $\mathcal{H}_2$ and $\mathcal{H}_\infty$ norms of the system were proposed. Both techniques were experimentally validated on a simply-supported piezoelectric laminate beam.

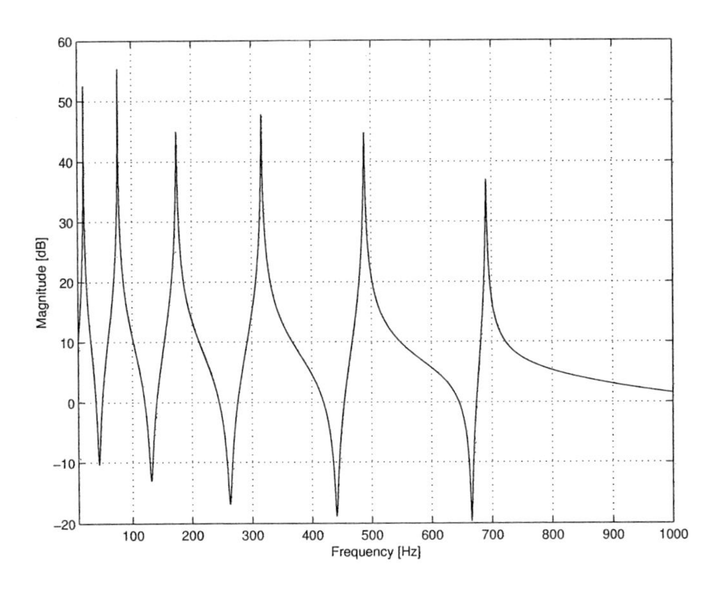

(a) magnitude

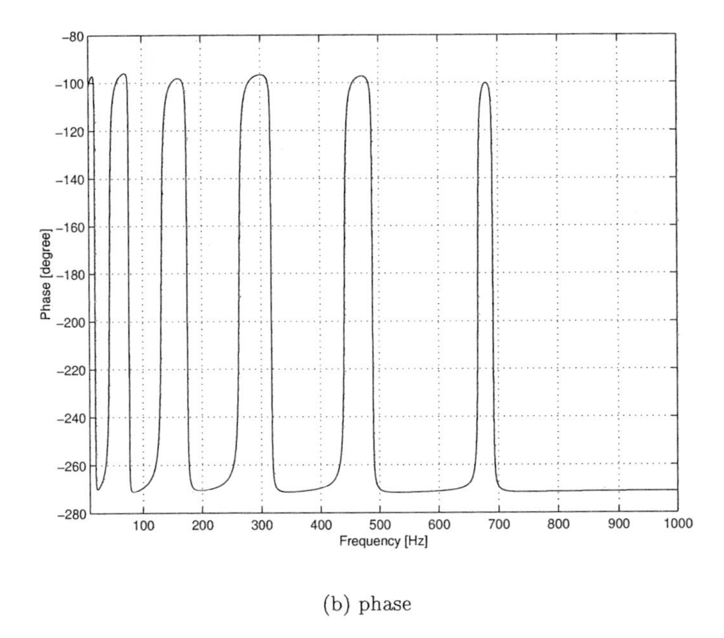

(b) phase

Fig. 5.17 Frequency response of the controller (input voltage to output voltage [V/V])

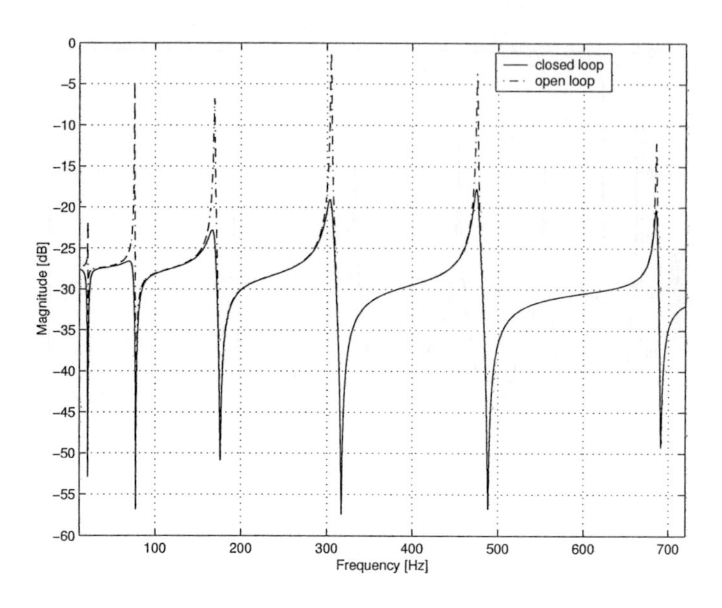

(a) simulation

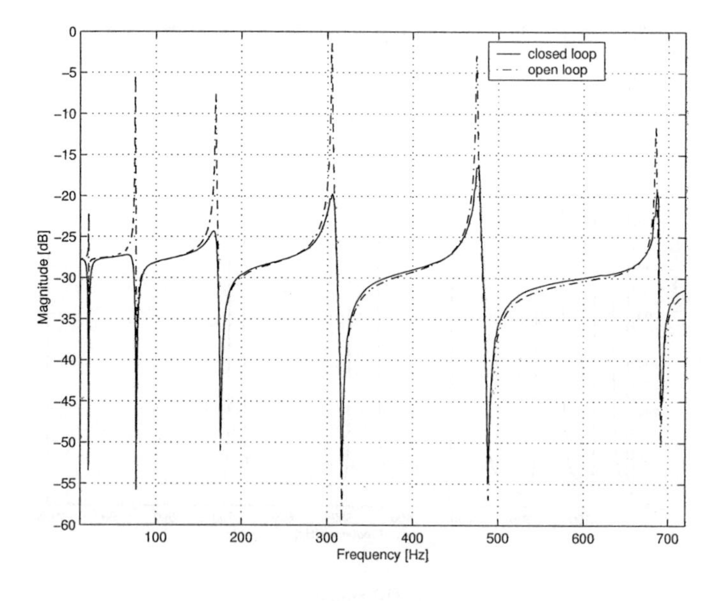

(b) experiment

Fig. 5.18 Simulation and experimental frequency responses (actuator voltage to sensor voltage [V/V]) - Magnitude

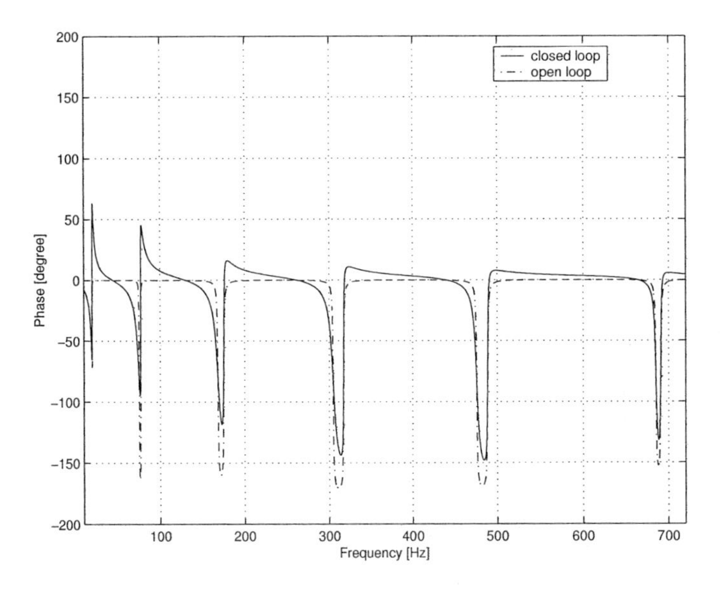

(a) simulation

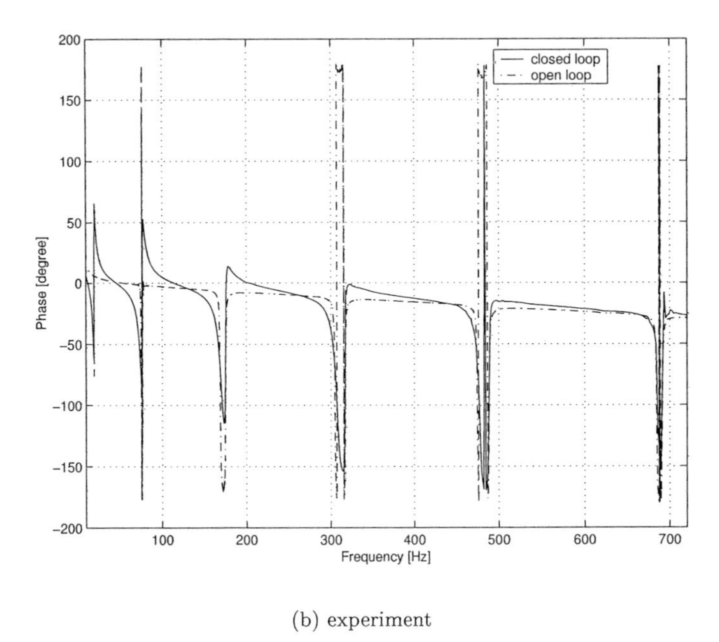

(b) experiment

Fig. 5.19 Simulation and experimental frequency responses (actuator voltage to sensor voltage [V/V]) - Phase

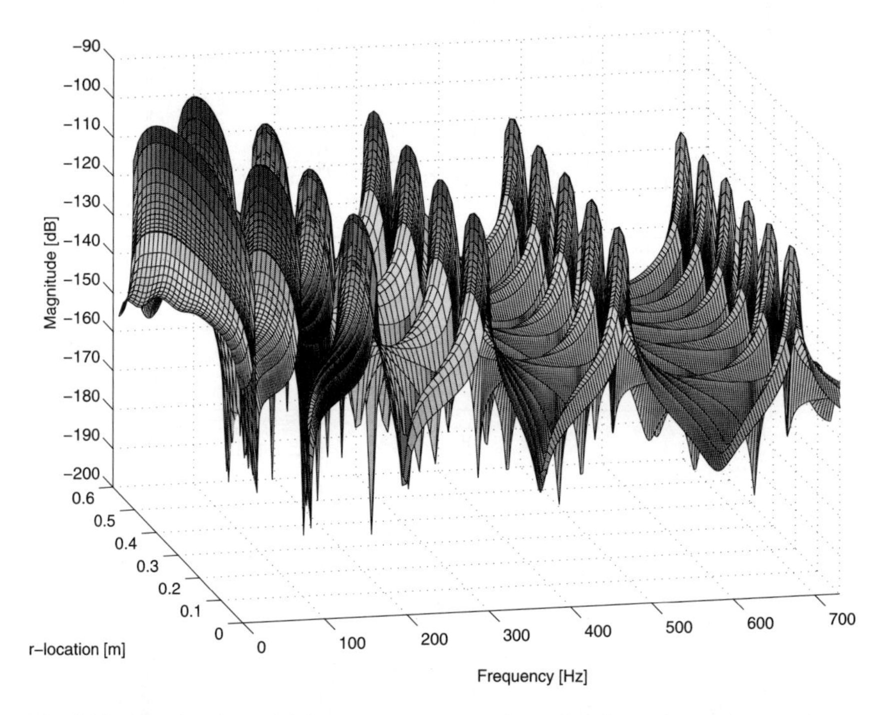

Fig. 5.20 Simulated spatial frequency response: (open loop) actuator voltage to beam deflection [m/V]

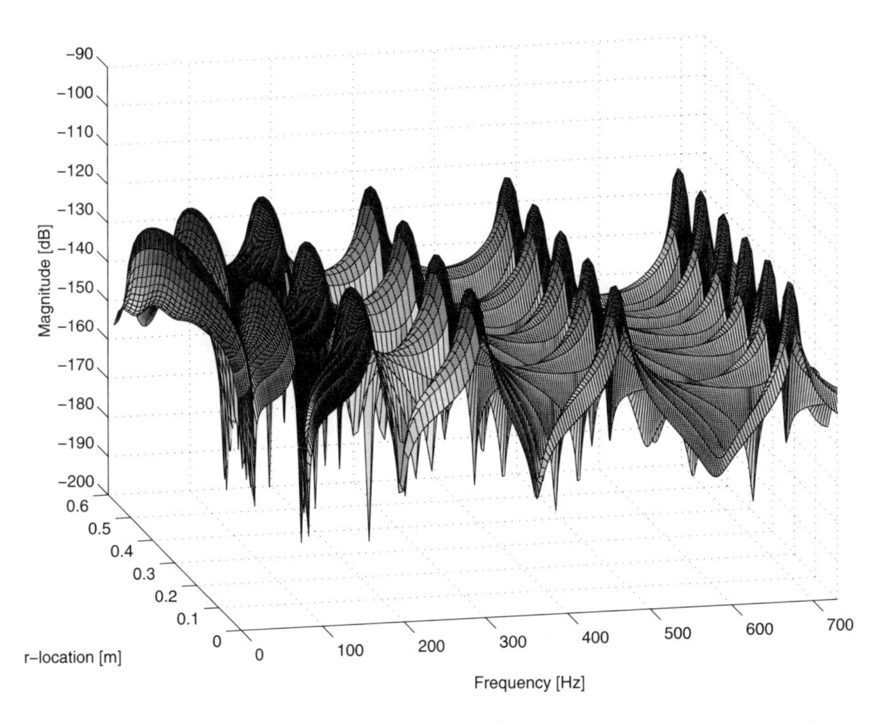

Fig. 5.21 Simulated spatial frequency response: (closed loop) actuator voltage to beam deflection [m/V]

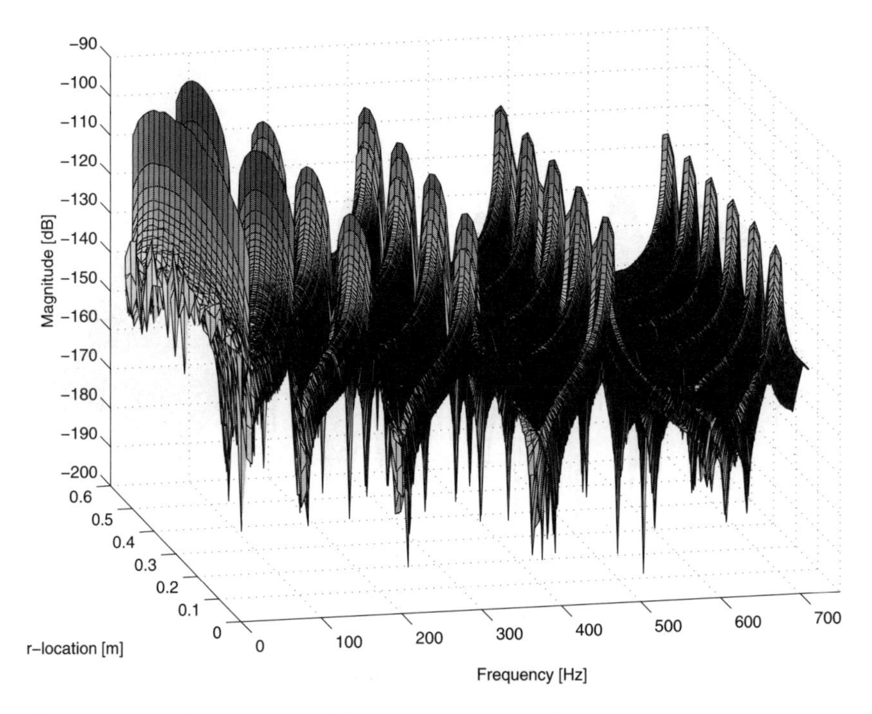

Fig. 5.22 Experimental spatial frequency response: (open loop) actuator voltage to beam deflection [m/V]

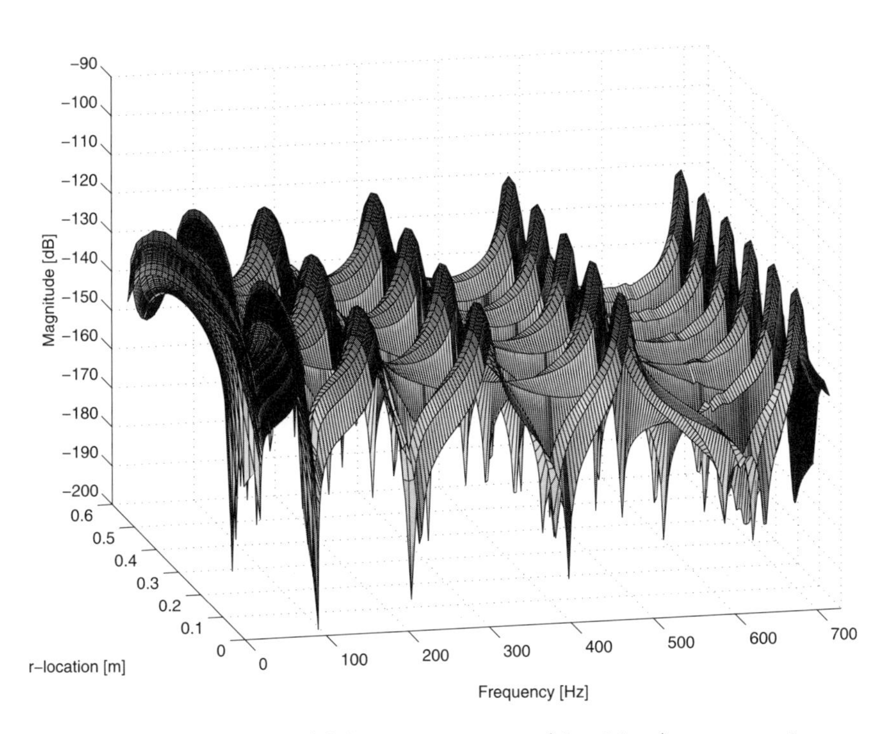

Fig. 5.23 Experimental spatial frequency response: (closed loop) actuator voltage to beam deflection [m/V]

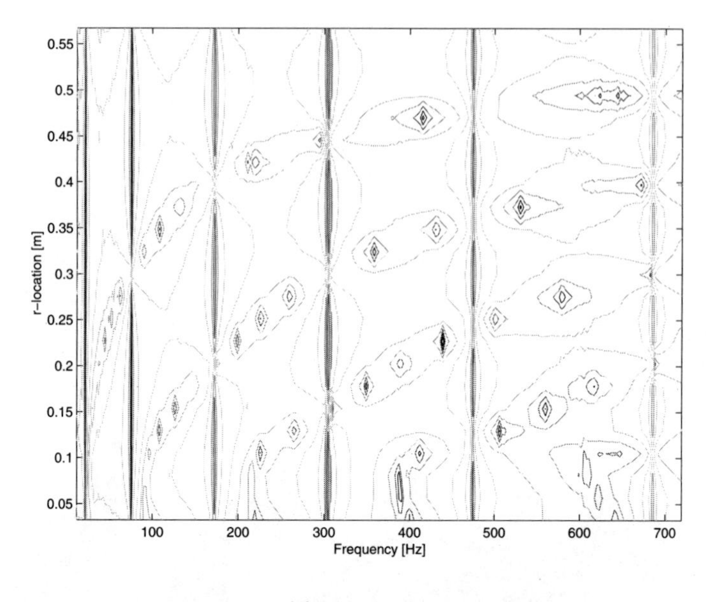

(a) open loop

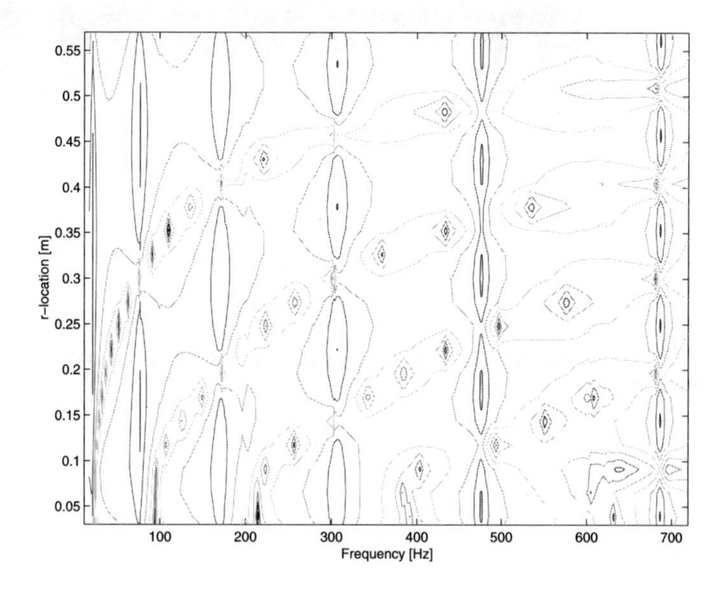

(b) closed loop

Fig. 5.24 Contour plots from experiments

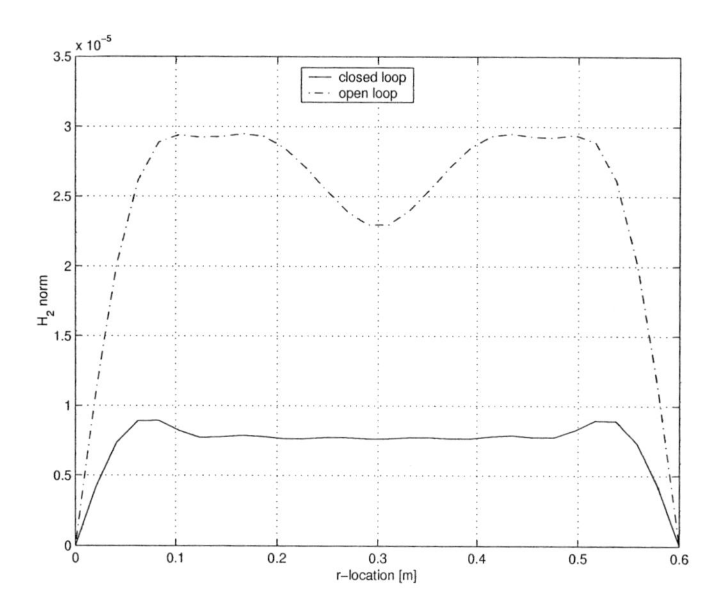

(a) spatial control

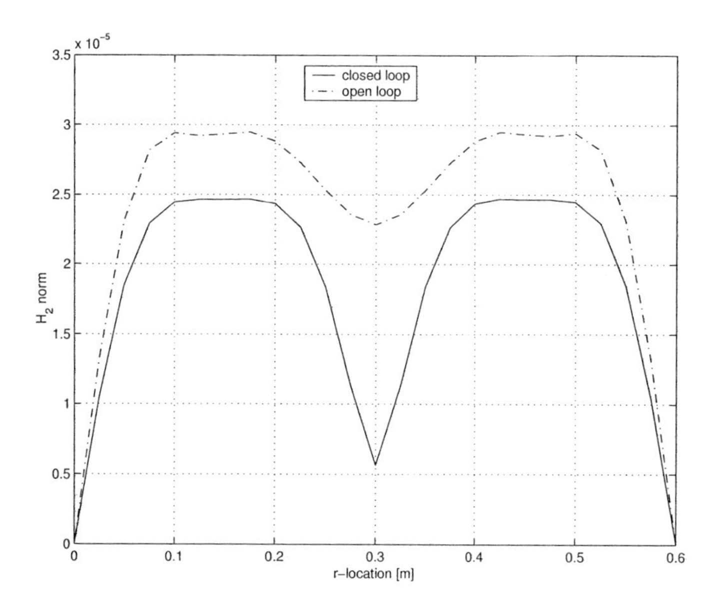

(b) point-wise control

Fig. 5.25 Simulated $\mathcal{H}_2$ norm plots

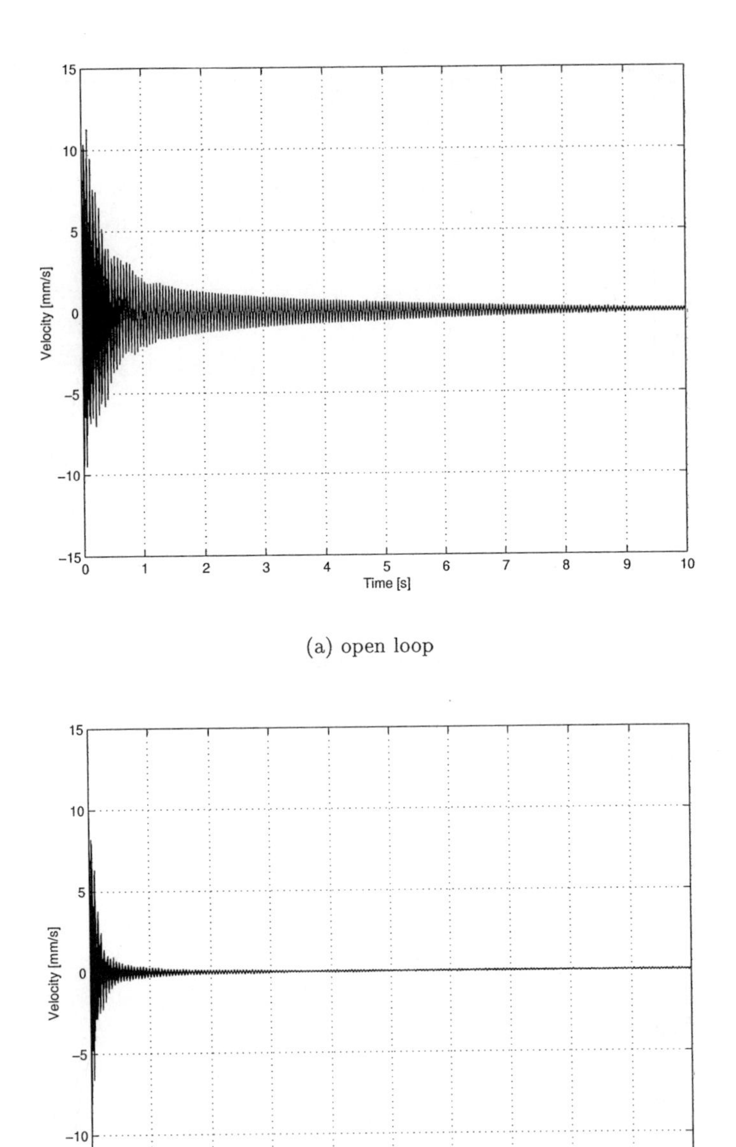

(a) open loop

(b) closed loop

Fig. 5.26 Vibration at the middle of the beam

Chapter 6

Optimal Placement of Actuators and Sensors

6.1 Introduction

An important aspect of controller design for spatially distributed systems is the locations at which actuators and sensors are mounted on the system. For control design purposes, it is desirable to place actuators and sensors at locations where one has adequate authority to control and/or observe specific modes of the system. This raises the issue of the need for meaningful measures of controllability and observability for such systems. This chapter is aimed at addressing this specific issue as well as developing a framework for optimal placement of actuators and sensors on flexible structures for vibration control purposes.

Due to its importance, the problem of optimal placement of actuators and sensors has been addressed by a number of researchers in the past. For example, references [28, 27, 98] address this problem using the notion of Hankel singular values of a system. Of particular recent interest is the placement of piezoelectric actuators and sensors on flexible structures. Reference [14] suggests that a piezoelectric actuator should be placed at a location of high average strain of the desired modes. Furthermore, authors of [43] place a collocated piezoelectric actuator/sensor pair on an all-clamped thin plate by identifying the location of the highest position sensitivity of each mode. Other researchers (see, for example, [20, 17]) use the optimization of quadratic performance indices to find optimal location for piezoelectric actuators and sensors. This approach to optimal placement, however, is dependent on the choice of the control law.

In this chapter, the problem of optimal placement of actuators and sensors is approached via our definitions of modal and spatial controllability and observability, where the latter definition is intimately related to the notion of spatial $\mathcal{H}_2$ norm defined earlier in the book. This allows us to

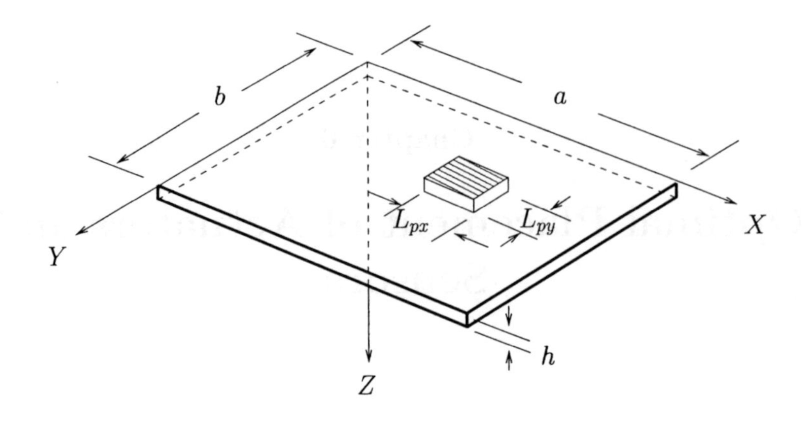

Fig. 6.1 A plate with the jth piezoelectric patch attached

study the problem from a meaningful perspective, and suggest placement strategies regardless of the nature of control techniques that are to be used.

The actuator and sensor placement methodology proposed in this chapter is validated on a flexible plate with simply-supported boundary conditions, and with a pair of collocated piezoelectric transducers. The next section is concerned with studying the dynamics of such a system. This will be followed by introducing our placement technique as well as its validation on an experimental rig.

6.2 Dynamics of a piezoelectric laminate plate

Consider a thin plate with dimensions of $a \times b \times h$ as shown in Figure 6.1. In Chapter 2, the partial differential equation that describes transverse vibration of a thin uniform plate was shown to be

$$\rho h \frac{\partial^2 w}{\partial t^2} + D\nabla^4 w(x,y,t) = \frac{\partial^2 M_{px}}{\partial x^2} + \frac{\partial^2 M_{py}}{\partial y^2}. \tag{6.1}$$

All parameters were defined in Chapter 2.

Let us assume that there are J actuators distributed over the plate. All piezoelectric patches are rectangular and have similar orientations. The right hand side of (6.1) represents the external moment per unit length generated by piezoelectric actuators. This, in turn, is directly related to the voltages applied to the piezoelectric transducers. The forcing term contributed by all actuators can be represented by $\sum_{j=1}^{J} \alpha_j V_{aj}(t)$, where $V_{aj}(t)$ is the voltage applied to the jth piezoelectric actuator. The term α_j

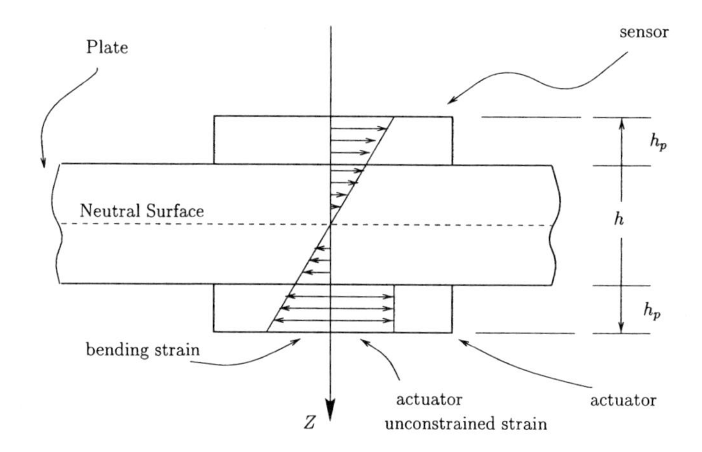

Fig. 6.2 A plate section with piezoelectric patches attached

depends on the properties and location of the jth actuator.

The derivation of a model for this system is along the same lines as that of a beam, but now it is extended to two-dimensional structures [18, 25]. Let us assume that the jth piezoelectric actuator is of dimensions $L_{px} \times L_{py} \times h_p$, as shown in Figure 6.1. To proceed, we need to make an assumption on the dimensions of the patches compared to that of the base structure. In particular, the actuator patches are assumed to be much thinner than the plate, so their contributions to structural properties can be neglected. That is, the mass and stiffness contributions of the patches are ignored. This is a reasonable assumption since many piezoelectric transducers commonly used in vibration control applications satisfy this condition, i.e. the patches are significantly thinner than the base structure.

Now, let us consider the bending deformation of a plate structure. The longitudinal stresses acting inside the actuator in the X and Y directions, σ_{px} and σ_{py}, are contributed by the bending strain, ϵ_x and ϵ_y, and the unconstrained strain, ϵ_{px} and ϵ_{py} (see Figure 6.2).

The unconstrained strains due to the applied voltage, $V_{aj}(t)$, are [25, 3]

$$\begin{aligned} \epsilon_{px} &= \left(\frac{d_{31}}{h_p}\right) V_{aj}(t) \\ \epsilon_{py} &= \left(\frac{d_{32}}{h_p}\right) V_{aj}(t). \end{aligned} \tag{6.2}$$

The piezoelectric charge constants along X and Y axes are denoted by d_{31} and d_{32} respectively. These constants dictate how much mechanical strain is generated when a particular potential difference is established across the piezoelectric patch.

Hooke's Law for plane stress is next used to obtain the longitudinal stresses in the X and Y directions, i.e. σ_{px} and σ_{py} respectively:

$$\begin{aligned} \sigma_{px} &= \frac{E_p}{1-\nu_p^2}\left(\epsilon_x - \epsilon_{px} + \nu(\epsilon_y - \epsilon_{py})\right) \\ \sigma_{py} &= \frac{E_p}{1-\nu_p^2}\left(\epsilon_y - \epsilon_{py} + \nu(\epsilon_x - \epsilon_{px})\right), \end{aligned} \tag{6.3}$$

where E_p and ν_p represent the Young's modulus of elasticity and Poisson's ratio of the actuator respectively. Also, the longitudinal strains in the X and Y directions are denoted by ϵ_x and ϵ_y.

The longitudinal stresses inside the structure, σ_x and σ_y also follow from the Hooke's Law:

$$\begin{aligned} \sigma_x &= \frac{E}{1-\nu^2}\left(\epsilon_x + \nu\epsilon_y\right) \\ \sigma_y &= \frac{E}{1-\nu^2}\left(\epsilon_y + \nu\epsilon_x\right). \end{aligned} \tag{6.4}$$

For pure bending case, the strain distribution across the thickness of the plate can be assumed to be linear as shown in Figure 6.2. The strain can be defined as $\epsilon_x = \alpha_x\, z$ and $\epsilon_y = \alpha_y\, z$, where α_x and α_y are the strain gradients and z is the transverse distance from the neutral surface of the plate (see Figure 6.2). These strain gradients can be determined from the moment equilibrium equations below. This equilibrium condition assumes a perfect bonding between the piezoelectric patch and the plate:

$$\begin{aligned} \int_{-\frac{h}{2}}^{\frac{h}{2}} z\,\sigma_x\, dz + \int_{\frac{h}{2}}^{\frac{h}{2}+h_p} z\,\sigma_{px}\, dz = 0 \\ \int_{-\frac{h}{2}}^{\frac{h}{2}} z\,\sigma_y\, dz + \int_{\frac{h}{2}}^{\frac{h}{2}+h_p} z\,\sigma_{py}\, dz = 0. \end{aligned} \tag{6.5}$$

The strain gradients are related to the unconstrained strains following the moment equilibrium equations (6.5):

$$\begin{aligned} \alpha_x &= \kappa\,\epsilon_{px} \\ \alpha_y &= \kappa\,\epsilon_{py}, \end{aligned} \tag{6.6}$$

where

$$\kappa = \frac{12\,E_p\,h_p\,(h_p + h)}{24\,D\,(1-\nu_p^2) + E_p\,\left[(h+2\,h_p)^3 - h^3\right]}. \tag{6.7}$$

Suppose that the ends of the jth actuator patch are located at (x_{1j}, y_{1j}) and (x_{2j}, y_{2j}). The external moment per unit length generated by this actuator can be obtained from the first integral term in (6.5) using (6.6). We consider the case where $d_{31} = d_{32}$ and we employ a step function $H(\cdot)$ to represent the spatial placement of the actuator:

$$M_{pxj} = M_{pyj} = \bar{A}_j \left[H(x - x_{1j}) - H(x - x_{2j})\right] \times \\ \left[H(y - y_{1j}) - H(y - y_{2j})\right] V_{aj}(t), \tag{6.8}$$

where

$$\bar{A}_j = \frac{\kappa_j d_{31j} D(1+\nu)}{h_{pj}}. \tag{6.9}$$

The external moments generated by the actuators, $\partial^2 M_{pxj}/\partial x^2$ and $\partial^2 M_{pyj}/\partial x^2$, can be obtained by differentiating M_{pxj} and M_{pyj} expressions twice with respect to x and y respectively, followed by using some specific properties of the Dirac delta function [52].

The partial differential equation for the transverse vibration of piezoelectric laminate thin plates (6.1) can now be solved using the modal analysis technique assuming a solution of the form:

$$w(x,y,t) = \sum_{m=1}^{\infty} \sum_{n=1}^{\infty} \phi_{mn}(x,y)\, q_{mn}(t). \tag{6.10}$$

The eigenvalue problem associated with (6.1) can be written as

$$D\, \nabla^4 \phi_{mn} = \rho\, h\, \omega_{mn}^2 \phi_{mn}. \tag{6.11}$$

Here, ω_{mn} and ϕ_{mn} represent the natural frequency and eigenfunction associated with the mode (m,n) respectively. Since D and ρh are constants, the eigenfunction ϕ_{mn} can be shown to satisfy the following orthogonality properties:

$$\rho h \int_0^b \int_0^a \phi_{mn}\, \phi_{pq}\, dx\, dy = \delta_{mp}\, \delta_{nq}$$
$$D \int_0^b \int_0^a \nabla^4 \phi_{mn}\, \phi_{pq}\, dx\, dy = \omega_{mn}^2\, \delta_{mp}\, \delta_{nq}, \tag{6.12}$$

where δ_{mp} and δ_{nq} are Kronecker delta functions.

Here, the contribution of external moments from all J actuators are considered. A set of uncoupled Ordinary Differential Equations can be obtained from the partial differential equation (6.1) using the orthogonality properties (6.12):

$$\omega_{mn}^2 q_{mn}(t) + 2\zeta_{mn}\omega_{mn}\dot{q}_{mn}(t) + \ddot{q}_{mn}(t) = \sum_{j=1}^{J} \int_0^b \int_0^a \phi_{mn}(x,y)\alpha_j dx\, dy\, V_{aj}(t), \tag{6.13}$$

where $q_{mn}(t)$ is the generalized coordinate and ζ_{mn} is the proportional damping term. Notice that the second term in (6.13) has been added to take into account the effect of damping, which is missing from (6.1).

Now, after using the Dirac delta function property (2.111), the right-hand-side of (6.13) can be shown to be

$$\sum_{j=1}^{J} \bar{A}_j \Psi_{mnj} V_{aj}(t),$$

where

$$\Psi_{mnj} = \left[\int_{y_{1j}}^{y_{2j}} \frac{\partial\phi_{mn}(x_{2j},y)}{\partial x} dy - \int_{y_{1j}}^{y_{2j}} \frac{\partial\phi_{mn}(x_{1j},y)}{\partial x} dy\right] + \left[\int_{x_{1j}}^{x_{2j}} \frac{\partial\phi_{mn}(x,y_{2j})}{\partial y} dx - \int_{x_{1j}}^{x_{2j}} \frac{\partial\phi_{mn}(x,y_{1j})}{\partial y} dx\right]. \tag{6.14}$$

Applying Laplace transform to (6.13) and assuming zero initial conditions, the transfer function from the voltages applied to the actuators,

$$V_a(s) = [V_{a1}(s), \ldots, V_{aJ}(s)]'$$

to the deflection of the plate $w(s,x,y)$ is found to be

$$G(s,x,y) = \sum_{m=1}^{\infty}\sum_{n=1}^{\infty} \frac{\phi_{mn}(x,y)\, P_{mn}}{\omega_{mn}^2 + 2\zeta_{mn}\omega_{mn} s + s^2}, \tag{6.15}$$

where

$$P_{mn} = \left[\int_0^b \int_0^a \phi_{mn}(x,y)\alpha_1 dxdy, \ldots, \int_0^b \int_0^a \phi_{mn}(x,y)\alpha_J dxdy\right], \tag{6.16}$$

that is,

$$P_{mn} = \left[\bar{A}_1 \Psi_{mn1}, \ldots, \bar{A}_J \Psi_{mnJ}\right]. \tag{6.17}$$

The transfer function (6.15) describes the dynamics of the structure when excited by piezoelectric actuators. To measure structural vibration for feedback control, some form of sensor is needed. Different types of sensors are available for this purpose. Piezoelectric sensors are commonly used for smart structures applications due to their low cost, light weight, availability, good sensing capabilities, and the capacity to be used as actuators simultaneously (i.e. self-sensing actuators), to mention a few.

Suppose there are J sensors distributed over the structure. A general description for the jth sensor output is given by

$$v_j(t) = \sum_{m=1}^{\infty} \sum_{n=1}^{\infty} \left(B_{mnj}\, q_{mn}(t) + \bar{B}_{mnj}\, \dot{q}_{mn}(t) \right), \tag{6.18}$$

where B_{mnj} and $\bar{B}_{mnj}$ depend on the properties and location of the sensor.

The sensing mechanism associated with the piezoelectric transducers is discussed next (see also [25, 57]). When a piezoelectric sensor is mechanically strained due to the structural deformation, it produces electric charges and consequently a voltage. The amount of voltage generated by the patch depends on the amount of deformation experienced by the base structure.

Let us consider the jth piezoelectric sensor attached to the plate. The electric charge generated inside the patch is related to the strain according to:

$$q(t) = \frac{k_{31}^2}{g_{31}}\, \epsilon_x + \frac{k_{32}^2}{g_{32}}\, \epsilon_y, \tag{6.19}$$

where k_{31} and k_{32} are the electromechanical coupling factors in X and Y directions respectively. Also, g_{31} and g_{32} are the voltage constants in X and Y directions. It is assumed that the electric charge generated inside the piezoelectric patch is caused entirely by the strains in X and Y directions, while the contribution of the shear strain to the total electric charge has been ignored. This is a valid assumption since the sensor's axes coincide with the geometrical axes of the plate, i.e. the sensor and the plate have similar orientation.

Assuming that the piezoelectric sensor was placed on the top surface of the plate (see Figure 6.2), an expression for the strains along X and Y directions in terms of the plate deflections can be obtained. In this case, z is the average distance from the neutral surface of the plate to the mid-plane of the sensor patch, i.e. $z = -(h + h_p)/2$,

$$\epsilon_x = \frac{\partial u}{\partial x} = \frac{h + h_p}{2} \frac{\partial^2 w}{\partial x^2}$$
$$\epsilon_y = \frac{\partial v}{\partial y} = \frac{h + h_p}{2} \frac{\partial^2 w}{\partial y^2}. \tag{6.20}$$

The induced voltage across the patch, $V_{sj}(t)$, can be obtained by realizing that the piezoelectric patch can be regarded as a parallel plate capacitor when it is charged.

Incorporating the strain expression in (6.20) into (6.19) and integrating the electric charge across the area of the sensor we obtain

$$V_{sj}(t) = \frac{h + h_{pj}}{2C_j} \int_{y_{1j}}^{y_{2j}} \int_{x_{1j}}^{x_{2j}} \left[\frac{k_{31j}^2}{g_{31j}} \frac{\partial^2 w}{\partial x^2} + \frac{k_{32j}^2}{g_{32j}} \frac{\partial^2 w}{\partial y^2} \right] dx\, dy, \tag{6.21}$$

where C_j is the piezoelectric capacitance of the jth patch.

After substituting the modal analysis solution (6.10), V_{sj} can be shown to be:

$$V_{sj}(t) =$$
$$\frac{h + h_{pj}}{2C_j} \sum_{m=1}^{\infty} \sum_{n=1}^{\infty} \left\{ \frac{k_{31j}^2}{g_{31j}} \left[\int_{y_{1j}}^{y_{2j}} \frac{\partial \phi_{mn}(x_{2j}, y)}{\partial x} dy - \int_{y_{1j}}^{y_{2j}} \frac{\partial \phi_{mn}(x_{1j}, y)}{\partial x} dy \right] \right.$$
$$\left. + \frac{k_{32j}^2}{g_{32j}} \left[\int_{x_{1j}}^{x_{2j}} \frac{\partial \phi_{mn}(x, y_{2j})}{\partial y} dx - \int_{x_{1j}}^{x_{2j}} \frac{\partial \phi_{mn}(x, y_{1j})}{\partial y} dx \right] \right\} q_{mn}(t). \tag{6.22}$$

It should be clear that the voltage generated by a piezoelectric sensor depends on the location at which the sensor is mounted on the structure as well as the eigenfunction associated with each mode. Thus, the placement of a sensor is crucial in determining the sensor's sensitivity in observing a certain number of modes.

6.3 Optimal placement of actuators

In this section we propose an optimization methodology for actuator placement based on the concept of spatial $\mathcal{H}_2$ norm of a spatially distributed system. The idea is to place actuators at locations where they can provide acceptable control authority over the structure in a spatial sense. The relevant spatial measures are defined in this section.

6.3.1 *Spatial controllability and modal controllability measures*

Consider a thin plate with J actuators, whose transfer function is given in (6.15). $G(s,x,y)$ describes the response of the plate at a point (x,y) due to the excitation generated by the actuators.

The spatial $\mathcal{H}_2$ norm of the transfer function of a plate, $G(s,x,y)$ in (6.15), can be calculated by taking advantage of the orthonormality properties of the eigenfunctions in (6.12) [78, 76], that is,

$$\begin{aligned} \ll G \gg_2^2 &= \frac{1}{2\pi} \int_{-\infty}^{\infty} \int_0^a \int_0^b \operatorname{tr}\{G(j\omega,x,y)^* G(j\omega,x,y)\} dy\, dx\, d\omega \\ &= \sum_{m=1}^{\infty} \sum_{n=1}^{\infty} \|\widetilde{G}_{mn}\|_2^2, \end{aligned} \tag{6.23}$$

where

$$\widetilde{G}_{mn} = \frac{P_{mn}}{\sqrt{\rho h}(s^2 + 2\zeta_{mn}\omega_{mn}s + \omega_{mn}^2)}. \tag{6.24}$$

For the more general case of a non-uniform plate structure, a model can be obtained using a numerical method such as the finite element technique [9, 84, 93, 10]. In this case, the spatial $\mathcal{H}_2$ norm of the transfer function can be calculated numerically. An alternative to this would be to use the spatial system identification technique which is explained in the next chapter.

Notice that $\widetilde{G}_{mn}(s)$, as described in (6.24), is the contribution of the mode (m,n) to the spatial $\mathcal{H}_2$ norm of $G(s,x,y)$. This is a nice property that results from orthonormality of mode shapes. This property also gives an indication of how much control authority the actuators have over the structure for each specific mode. It can be observed that the level of control authority depends on the location at which the actuators are placed on the structure.

It should be noted that the above result is different from the additive property of modal norms used in reference [28]. The result in (6.23) arises from the spatial $\mathcal{H}_2$ norm definition for the transfer function which involves a spatial averaging on the spatial model, while the additive property in [28] is an approximation based on the fact that the cross-coupling between modes is small for lightly damped flexible structures and is not based on a spatial averaging on the system.

Since $\|\widetilde{G}_{mn}\|_2$ in (6.23) describes the level of authority of actuators over each mode, this measure can be used to determine the effectiveness of actuators in controlling each mode. We define a function:

$$f_{mn}(x_1, y_1) = \|\widetilde{G}_{mn}\|_2, \tag{6.25}$$

where $f_{mn}(x_1, y_1)$ implies that f_{mn} is a function of the locations at which the actuators are mounted on the structure. The function can be normalized with respect to its maximum value. This normalized value is then referred to as the *modal controllability* [78]:

$$\mathcal{M}_{mn}(x_1, y_1) = \frac{f_{mn}(x_1, y_1)}{\alpha_{mn}} \times 100\%, \tag{6.26}$$

where

$$\alpha_{mn} = \max_{(x_1, y_1) \in \mathcal{R}_1} f_{mn}(x_1, y_1)$$

and $\mathcal{R}_1 \subset \mathcal{R}$. For the case of a thin plate of dimensions $a \times b$,

$$\mathcal{R} = \{(x, y) \mid 0 \leq x \leq a, 0 \leq y \leq b\}$$

and $\mathcal{R}_1$ is the set of all possible locations where the actuators can be mounted on the structure.

If an actuator is meant to have high authority over a specific mode, it should be placed at a location where the modal controllability associated with that mode is large. On the other hand, if it is not desirable to excite a particular mode significantly, the actuator needs to be placed at a location where the modal controllability of that mode is low.

The spatial $\mathcal{H}_2$ norm of the system in (6.23) describes the control authority of the actuators over the entire structure in a spatially averaged sense. Placement of the actuators on the structure would obviously affect this level of authority. From (6.23) we can observe that the contribution of high-frequency modes to the spatial $\mathcal{H}_2$ norm of the system tends to decrease with frequency. The low-frequency modes tend to contribute more to the vibrations of structures as expected, although the amount of contribution also depends on where the actuators are located. Therefore, it is reasonable to consider only a number of low-frequency modes to determine the system's spatial $\mathcal{H}_2$ norm.

If only the I_m lowest frequency modes are taken into account, the *spatial controllability* is defined as follows:

$$\mathcal{S}_c(x_1, y_1) = \frac{1}{\beta}\sqrt{\sum_{i=1}^{I_m} f_{m_i n_i}(x_1, y_1)^2} \times 100\%, \tag{6.27}$$

where

$$\beta = \max_{(x_1,y_1)\in\mathcal{R}_1} \sqrt{\sum_{i=1}^{I_m} f_{m_i n_i}(x_1,y_1)^2}.$$

Here, m_i and n_i correspond to mode (m_i, n_i), which is the ith lowest frequency mode of the system. Note that spatial controllability is the spatial $\mathcal{H}_2$ norm of the system calculated based on a limited number of modes and normalized with respect to β.

Hence, the modal controllability is a measure of controller authority over each mode, while the spatial controllability represents the actuator authority contributed by selected modes. A suitable location for the actuators can be determined by maximizing the spatial controllability measure. However, placing actuators in the location of highest spatial controllability may not be enough to ensure good performance for every mode. It is possible that at the chosen locations, modal controllability of several modes may be unacceptably low. As a consequence, it is also important to maintain a reasonable level of actuator control authority over each mode. One can thus set the minimum level for the modal controllability associated with each mode as an additional constraint in the optimization. This would guarantee a minimum level of modal controllability for each mode, as well as an acceptable spatial controllability for the entire structure.

An optimization problem can now be set up. The actuators are to be placed at locations where they provide sufficiently high spatial controllability. However, at the same time, the modal controllability of some important modes have to be above certain levels. The constrained optimization problem for the actuator placement is

$$\begin{aligned} &\max_{(x_1,y_1)\in\mathcal{R}_1} \quad \mathcal{S}_c(x_1,y_1) \\ &\text{subject to: } \mathcal{M}_{m_i n_i}(x_1,y_1) \geq b_i,\ i = 1,2,\ldots,I_m, \end{aligned} \tag{6.28}$$

where (m_i, n_i) is the mode corresponding to the ith lowest frequency mode, and b_i is the lowest allowable level for modal controllability of mode (m_i, n_i).

6.3.2 *Control spillover reduction*

Let us assume that the actuators are "optimally" placed according to the procedure explained in the previous subsection. This would mean that the actuators should provide high spatial controllability and sufficient modal controllability for the targeted low-frequency modes of the system. A problem that may arise is that the actuators would, inadvertently, provide high authority over a number of high-frequency modes too. This situation can contribute to control spillover which may cause instability or loss of performance when a feedback controller is implemented on the system due to the actuators exciting the higher frequency modes [48]. It is thus important to be able to place the actuators at locations where they have less authority over the high-frequency modes. This amounts to ensuring that those specific modes do not enjoy a high level of modal controllability.

To reduce the control spillover effect, we can add extra modal controllability constraints that limit the authority of the actuator over some high-frequency modes [33]. As a consequence, more constraints need to be added to the optimization problem. Another alternative is to add only one extra constraint that would guarantee a sufficiently low level of spatial controllability for several higher frequency modes. By doing so, the spillover effect on the system can be reduced. Several higher frequency modes can be targeted and the spatial controllability associated with these modes, $\mathcal{S}_{c2}$, can be expressed as

$$\mathcal{S}_{c2}(x_1, y_1) = \frac{1}{\beta_2}\sqrt{\sum_{i=I_m+1}^{\bar{I}} f_{m_i n_i}(x_1, y_1)^2} \times 100\%, \tag{6.29}$$

where

$$\beta_2 = \max_{(x_1,y_1)\in\mathcal{R}_1} \sqrt{\sum_{i=I_m+1}^{\bar{I}} f_{m_i n_i}(x_1, y_1)^2}$$

and $\bar{I}$ corresponds to the highest frequency mode that is targeted for the control spillover reduction. Therefore, the optimization problem becomes

$$\begin{aligned} &\max_{(x_1,y_1)\in\mathcal{R}_1} \quad \mathcal{S}_c(x_1, y_1) \\ &\text{subject to: } \mathcal{M}_{m_i n_i}(x_1, y_1) \geq b_i,\ i = 1, 2, \ldots, I_m, \\ &\qquad\qquad\quad \mathcal{S}_{c2}(x_1, y_1) \leq c \end{aligned} \tag{6.30}$$

where c is the highest allowable level of spatial controllability for spillover reduction.

We can set the allowable limits for modal controllability and spatial controllability depending on the practical requirements. For reduction of control spillover, it is important to include only a few dominant higher frequency modes. The more high-frequency modes included in $\mathcal{S}_{c2}$, the less freedom we may have in determining the upper level of the constraint, c. The reason is that we can always choose a location where modal controllability of one mode is reduced, but that of another mode is increased. This situation will worsen if more modes are included for the control spillover reduction. Hence, it is not practical to consider too many high-frequency modes. It may be sufficient to consider only a few such modes since the controller can be designed so that its response will roll off at higher frequencies. Thus, the spillover effect can be reduced even further.

An extension of this optimal placement methodology is to include only a number of selected modes that are of control significance, and not all of the low-frequency modes. In this case, the spatial controllability $\mathcal{S}_c$ is calculated based on only the selected modes, while $\mathcal{S}_{c2}$ is calculated based on several other modes that are important for spillover reduction purposes.

6.4 Optimal placement of sensors

The previous section dealt with the optimal placement of actuators based on the concept of spatial $\mathcal{H}_2$ norm of spatial systems. One may ask whether the concept can be extended to obtain a measure that describes the observation authority of sensors over the entire structure in a spatial sense. This section discusses this issue and develops an optimization methodology for optimal placement of sensors.

6.4.1 *Spatial and modal observability measures*

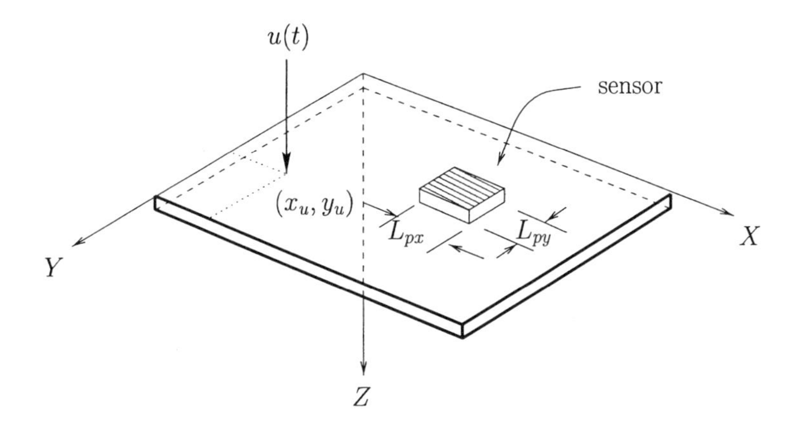

Fig. 6.3 A plate with a piezoelectric sensor and a point disturbance $u(t)$

Consider a point disturbance $u(t)$ acting on a structure at an arbitrary location (x_u, y_u), as shown in Figure 6.3. This disturbance will excite structural vibration that may be observed by sensors. The generalized force from the modal analysis in (6.13) is described as follows:

$$\int_0^b \int_0^a \phi_{mn}(x,y)\delta(x-x_u)\delta(y-y_u)dx\,dy\,u(t) = \phi_{mn}(x_u,y_u)u(t). \tag{6.31}$$

Suppose there are J sensors distributed over the structure. The jth sensor observes the structural vibration according to (6.18). The transfer function from the point disturbance $u(t)$ to the sensor output, $v(t) = [v_1(t), \ldots, v_J(t)]'$ is

$$G_{vu}(s, x_u, y_u) = \sum_{m=1}^{\infty}\sum_{n=1}^{\infty} \frac{\phi_{mn}(x_u,y_u)\,(Q_{mn} + \bar{Q}_{mn}s)}{s^2 + 2\zeta_{mn}\,\omega_{mn}\,s + \omega_{mn}^2}, \tag{6.32}$$

where

$$\begin{aligned} Q_{mn} &= [B_{mn1}, \ldots, B_{mnJ}]' \\ \bar{Q}_{mn} &= \left[\bar{B}_{mn1}, \ldots, \bar{B}_{mnJ}\right]'. \end{aligned} \tag{6.33}$$

Notice that the transfer function G_{vu} depends on where the disturbance is acting on the structure, i.e. it is a function of the spatial location (x_u, y_u).

If we wish to find the best location for the sensors, we may choose a place that minimizes energy of the signals detected by the sensors which is due to the point disturbance. A measure of the output energy due to a white noise input signal is represented by the $\mathcal{H}_2$ norm of the system, G_{vu}. This allows the sensor to be located at a place where it can observe effectively the structural vibration caused by a point disturbance at (x_u, y_u). However, the sensor may not be able to observe vibration caused by disturbances at other locations. To ensure that vibration caused by disturbances at other points can be observed in a spatial sense, it is necessary to include the spatial characteristics of the system into the optimization problem.

In order to incorporate the spatial information embedded in the system, we may use the spatial $\mathcal{H}_2$ norm as a measure of performance:

$$
\begin{aligned}
&\ll G_{vu} \gg_2^2 \\
&= \frac{1}{2\pi} \int_{-\infty}^{\infty} \int_0^a \int_0^b \left\{ G_{vu}(j\omega, x_u, y_u)^* G_{vu}(j\omega, x_u, y_u) \right\} dy_u \, dx_u \, d\omega \\
&= \sum_{m=1}^{\infty} \sum_{n=1}^{\infty} \bar{f}_{mn}^2, \qquad (6.34)
\end{aligned}
$$

where

$$
\bar{f}_{mn}(x_1, y_1) = \sqrt{\frac{1}{\rho h} \sum_{j=1}^{J} \left\| \frac{B_{mnj} + \bar{B}_{mnj} s}{s^2 + 2\zeta_{mn}\omega_{mn} s + \omega_{mn}^2} \right\|_2^2}. \qquad (6.35)
$$

Here, $\bar{f}_{mn}$ is a function of the locations of actuators that are expressed by (x_1, y_1). Note that G_{vu} is a column vector so no trace operation is required to compute the norm.

This measure differs from that associated with actuator placement problem as the input signal is a function of spatial locations of the structure. It can be observed that the spatial $\mathcal{H}_2$ norm of G_{vu} contains an independent contribution from each mode. It is interesting to compare this with the spatial $\mathcal{H}_2$ norm of G in (6.23). Both results show the contribution of each mode to the spatial $\mathcal{H}_2$ norm without any coupling between them. The non-existence of mode couplings greatly simplifies the optimization since the contribution of each mode to the spatial $\mathcal{H}_2$ norm of the system can be clearly defined.

To clarify the physical meaning of this, we point out that the system spatial $\mathcal{H}_2$ norm measures the detected signal energy of the sensor due to excitation of a point disturbance at all possible locations over the structure in a spatially averaged sense. This measure can be used to determine how effective the location of the sensor is in sensing the structural vibration caused by external disturbances. Since the system is assumed to be linear, a general case of disturbances can be regarded as a linear superposition of point disturbances. This implies that finding a location for a sensor where the value of the spatial $\mathcal{H}_2$ norm is large would generally result in high sensitivity for arbitrary disturbances as well.

Now, a normalized value of (6.35) is

$$
\mathcal{K}_{mn}(x_1, y_1) = \frac{\bar{f}_{mn}(x_1, y_1)}{\bar{\alpha}_{mn}} \times 100\%, \qquad (6.36)
$$

where

$$\bar{\alpha}_{mn} = \max_{(x_1,y_1)\in\mathcal{R}_1} \bar{f}_{mn}(x_1,y_1). \tag{6.37}$$

$\mathcal{K}_{mn}$ indicates the *modal observability* of the mode (m,n) since it determines the level of contribution of the mode to the spatial $\mathcal{H}_2$ norm of the system.

The spatial observability can now be set up using the result in (6.34). If only the I_m lowest frequency modes are taken into account, the *spatial observability* is defined as follows:

$$\mathcal{S}_o(x_1,y_1) = \frac{1}{\bar{\beta}}\sqrt{\sum_{i=1}^{I_m} \bar{f}_{m_i n_i}(x_1,y_1)^2} \times 100\%, \tag{6.38}$$

where

$$\bar{\beta} = \max_{(x_1,y_1)\in\mathcal{R}_1} \sqrt{\sum_{i=1}^{I_m} \bar{f}_{m_i n_i}(x_1,y_1)^2}.$$

Here, m_i and n_i correspond to mode (m_i,n_i), which is the ith lowest frequency mode. The spatial observability is the normalized value of the approximate spatial $\mathcal{H}_2$ norm of G_{vu}, where $\bar{\beta}$ is the maximum value of the norm. An optimization can then be set up by maximizing the spatial observability within constraints determined by modal observabilities.

6.4.2 *Observation spillover reduction*

Similar to control spillover, observability spillover may also affect the performance and stability of the closed-loop system. The signals measured by the sensors may contain substantial high-frequency components that are not of control significance. Here, we develop a technique for the reduction of observation spillover in parallel to the procedure proposed in the previous subsection. In particular, several higher frequency modes can be chosen and the spatial observability for these modes, $\mathcal{S}_{o2}$, can be expressed as

$$\mathcal{S}_{o2}(x_1,y_1) = \frac{1}{\bar{\beta}_2}\sqrt{\sum_{i=I_m+1}^{\bar{I}} \bar{f}_{m_i n_i}(x_1,y_1)^2} \times 100\%, \tag{6.39}$$

where

$$\bar{\beta}_2 = \max_{(x_1,y_1)\in\mathcal{R}_1} \sqrt{\sum_{i=I_m+1}^{\bar{I}} \bar{f}_{m_i n_i}(x_1,y_1)^2}$$

and $\bar{I}$ corresponds to the highest frequency mode that is included for the observation spillover reduction.

Therefore, the optimization problem becomes

$$\begin{aligned} &\max_{(x_1,y_1)\in\mathcal{R}_1} \quad \mathcal{S}_o(x_1,y_1) \\ &\text{subject to:}\, \mathcal{K}_{m_i n_i}(x_1,y_1) \geq b_i,\ i = 1,2,\ldots,I_m, \\ &\qquad\qquad \mathcal{S}_{o2}(x_1,y_1) \leq c \end{aligned} \tag{6.40}$$

where b_i is the minimum level for modal observability of mode (m_i, n_i), while c is the highest allowable level of spatial observability for spillover reduction.

Similar to the case for actuator placement, it is also possible to include only a number of selected modes to determine $\mathcal{S}_o$,$\mathcal{S}_{o2}$ and $\mathcal{K}_{mn}$. Therefore, we need to concentrate only on selected modes that are of control significance.

6.5 Optimal placement of piezoelectric actuators and sensors

In this section we apply the optimal placement methodology developed in the earlier parts of this chapter to a piezoelectric laminate plate.

Although it is possible to find the optimal placement for several actuators and sensors simultaneously, the procedure may result in a complicated optimization problem when these actuators and sensors have large dimensions relative to the structure. In practice, it may be easier to find the optimal placement for an individual piezoelectric actuator/sensor pair at a time, since it is less straightforward to set geometric constraints that prevent patches from overlapping each other during the optimization process. This is the approach used in this section.

6.5.1 *Piezoelectric actuators*

Let us consider the placement of a single actuator, say the jth actuator. In this case, we deal with a single input version of system $G(s, x, y)$ in (6.15), i.e. the transfer function from the jth actuator to the structural deflection of the plate. Based on this transfer function, the contribution of each mode (m,n) is $\|\widetilde{G}_{mnj}\|_2^2$ as in (6.24). Each mode's contribution depends on the location of the jth actuator on the structure. Thus, in order to find the optimal placement for the actuator, the contribution of mode (m,n) due to the jth actuator needs to be considered. That is, $\|\widetilde{G}_{mnj}\|_2^2$, where

$$\widetilde{G}_{mnj} = \frac{P_{mnj}}{\sqrt{\rho h}(s^2 + 2\zeta_{mn}\omega_{mn}s + \omega_{mn}^2)} \tag{6.41}$$

and $P_{mnj} = \bar{A}_j \Psi_{mnj}$ is described in (6.17). Suppose that one corner of the jth piezoelectric actuator patch with a fixed size is located at (x_{1j}, y_{1j}). A function f_{mnj} can be defined as

$$\begin{aligned} f_{mnj}(x_{1j}, y_{1j}) &= \|\widetilde{G}_{mnj}\|_2 \\ &= \left|\frac{\bar{A}_j \Psi_{mnj}(x_{1j}, y_{1j})}{\sqrt{\rho h}}\right| \left\|\frac{1}{s^2 + 2\zeta_{mn}s + \omega_{mn}^2}\right\|_2 . \end{aligned} \tag{6.42}$$

The modal controllability is

$$\mathcal{M}_{mn}(x_{1j}, y_{1j}) = \frac{f_{mnj}(x_{1j}, y_{1j})}{\alpha_{mnj}} \times 100\%, \tag{6.43}$$

where

$$\alpha_{mnj} = \max_{(x_{1j}, y_{1j}) \in \mathcal{R}_1} f_{mnj}(x_{1j}, y_{1j})$$

and $\mathcal{R}_1 \subset \mathcal{R}$. For the case of a thin plate of dimensions $a \times b$ and a piezoelectric actuator of dimensions $L_{px} \times L_{py}$,

$$\mathcal{R}_1 = \{(x, y) \mid 0 \le x \le a - L_{px}, 0 \le y \le b - L_{py}\}$$

and

$$\mathcal{R} = \{(x, y) \mid 0 \le x \le a, 0 \le y \le b\}.$$

If only the I_m lowest frequency modes are taken into account, the spatial controllability is defined as

$$\mathcal{S}_c(x_{1j}, y_{1j}) = \frac{1}{\beta_j}\sqrt{\sum_{i=1}^{I_m} f_{m_i n_i j}(x_{1j}, y_{1j})^2} \times 100\%, \tag{6.44}$$

where

$$\beta_j = \max_{(x_{1j}, y_{1j}) \in \mathcal{R}_1} \sqrt{\sum_{i=1}^{I_m} f_{m_i n_i j}(x_{1j}, y_{1j})^2}.$$

Here, m_i and n_i correspond to mode (m_i, n_i), which is the ith lowest frequency mode of the structure.

Several higher frequency modes can be chosen to reduce control spillover and the spatial controllability associated with these modes, $\mathcal{S}_{c2}$, which is expressed as

$$\mathcal{S}_{c2}(x_{1j}, y_{1j}) = \frac{1}{\beta_{2j}} \sqrt{\sum_{i=I_m+1}^{\bar{I}} f_{m_i n_i j}(x_{1j}, y_{1j})^2} \times 100\%, \tag{6.45}$$

where

$$\beta_{2j} = \max_{(x_{1j}, y_{1j}) \in \mathcal{R}_1} \sqrt{\sum_{i=I_m+1}^{\bar{I}} f_{m_i n_i j}(x_{1j}, y_{1j})^2}$$

and $\bar{I}$, again, corresponds to the highest frequency mode that is considered for the control spillover reduction. Then the optimization problem for the placement of piezoelectric actuators can be set up as in (6.30).

6.5.2 *Piezoelectric sensors*

We consider the case where $k_{31} = k_{32}$ and $g_{31} = g_{32}$. Substituting the modal analysis solution and the electric charge expression above, the voltage induced in the jth piezoelectric sensor $V_{sj}(t)$ in (6.22) can be shown to be:

$$V_{sj}(t) = \frac{k_{31j}^2}{C_j \, g_{31j}} \left(\frac{h + h_{pj}}{2} \right) \sum_{m=1}^{\infty} \sum_{n=1}^{\infty} \Psi_{mnj}(x_{1j}, y_{1j}) \, q_{mn}(t). \tag{6.46}$$

Following the description for general sensors, the modal observability associated with each mode can be obtained as

$$\mathcal{K}_{mn}(x_{1j}, y_{1j}) = \frac{\bar{f}_{mnj}(x_{1j}, y_{1j})}{\bar{\alpha}_{mnj}} \times 100\%, \tag{6.47}$$

where

$$\begin{aligned} \bar{f}_{mnj} &= \left| \frac{k_{31j}^2}{C_j \, g_{31j} \sqrt{\rho h}} \left(\frac{h + h_{pj}}{2} \right) \Psi_{mnj}(x_{1j}, y_{1j}) \right| \\ &\quad \times \left\| \frac{1}{s^2 + 2\zeta_{mn}\omega_{mn}s + \omega_{mn}^2} \right\|_2 \\ \bar{\alpha}_{mnj} &= \max_{(x_{1j}, y_{1j}) \in \mathcal{R}_1} \bar{f}_{mnj}(x_{1j}, y_{1j}). \end{aligned} \tag{6.48}$$

Similarly, the spatial observability is

$$\mathcal{S}_o(x_{1j}, y_{1j}) = \frac{1}{\bar{\beta}_j}\sqrt{\sum_{i=1}^{I_m} \bar{f}_{m_in_ij}(x_{1j}, y_{1j})^2} \times 100\%, \tag{6.49}$$

where

$$\bar{\beta}_j = \max_{(x_{1j},y_{1j})\in\mathcal{R}_1} \sqrt{\sum_{i=1}^{I_m} \bar{f}_{m_in_ij}(x_{1j}, y_{1j})^2}.$$

The spatial observability for observation spillover reduction is

$$\mathcal{S}_{o2}(x_{1j}, y_{1j}) = \frac{1}{\bar{\beta}_{2j}}\sqrt{\sum_{i=I_m+1}^{\bar{I}} \bar{f}_{m_in_ij}(x_{1j}, y_{1j})^2} \times 100\%, \tag{6.50}$$

where

$$\bar{\beta}_{2j} = \max_{(x_{1j},y_{1j})\in\mathcal{R}_1} \sqrt{\sum_{i=I_m+1}^{\bar{I}} \bar{f}_{m_in_ij}(x_{1j}, y_{1j})^2}.$$

The optimization problem for the placement of piezoelectric sensors can be set up as in (6.40).

Now, consider when identical and collocated piezoelectric patches are used as actuators and sensors. The piezoelectric transducers have similar properties in X and Y directions, i.e. $k_{31} = k_{32}$, $g_{31} = g_{32}$ and $d_{31} = d_{32}$. By comparing (6.43) and (6.47) we notice that $\mathcal{M}$ and $\mathcal{K}$ are clearly similar since they are both linearly proportional to $\Psi_{mnj}(x_{1j}, y_{1j})$. As a consequence, the associated spatial controllability and spatial observability are similar as well. That is,

$$\begin{aligned}\mathcal{M}_{mn}(x_{1j}, y_{1j}) &= \mathcal{K}_{mn}(x_{1j}, y_{1j}),\\ \mathcal{S}_c(x_{1j}, y_{1j}) &= \mathcal{S}_o(x_{1j}, y_{1j}) \text{ and}\\ \mathcal{S}_{c2}(x_{1j}, y_{1j}) &= \mathcal{S}_{o2}(x_{1j}, y_{1j}).\end{aligned} \tag{6.51}$$

Suppose we place a piezoelectric actuator on a structure, at a location that guarantees acceptable levels of spatial and modal controllabilities. Placing an identical piezoelectric sensor at the same location would yield similar levels of spatial and modal observabilities.

This result is beneficial for finding an optimal location for collocated piezoelectric actuator-sensor pairs since we need only find an appropriate location for either the actuators or the sensors. The following section discusses an illustrative example of the optimal placement of a collocated piezoelectric actuator-sensor pair on a thin rectangular plate.

6.6 Numerical and experimental results

In this section we consider the problem of optimal placement of a collocated piezoelectric actuator-sensor pair on a thin rectangular plate whose edges are pinned. Table 6.1 gives the properties of the piezoelectric laminate plate. Initially, each mode of the plate is assumed to have a damping ratio of $\zeta_{mn} = 0.002$. This, however, is later refined using the experimental data obtained from the rig.

Table 6.1 Properties of the piezoelectric laminate plate

Plate X-length, a	0.80 m
Plate Y-length, b	0.60 m
Plate thickness, h	0.004 m
Plate Young's Modulus, E	7.0×10^{10} N/m^2
Plate Poisson's ratio, ν	0.30
Plate density, ρh	11.0 kg/m^2
Piezoceramic X-length, L_{px}	0.0724 m
Piezoceramic Y-length, L_{py}	0.0724 m
Piezoceramic thickness, h_p	1.91×10^{-4} m
Piezoceramic Young's Modulus, E_p	6.20×10^{10} N/m^2
Piezoceramic Poisson's ratio, ν_p	0.30
Charge constant, d_{31}	-3.20×10^{-10} m/V
Voltage constant, g_{31}	-9.50×10^{-3} Vm/N
Capacitance, C	4.50×10^{-7} F
Electromechanical coupling factor, k_{31}	0.44

The optimization procedure described in the previous section is employed to determine an appropriate position for the placement of the actuator and sensor. The five lowest frequency modes are used to calculate the objective function $\mathcal{S}_c$. The next five modes are considered for the control spillover reduction. Table 6.2 shows the frequencies of the relevant modes based on the simulated model.

Figures 6.4, 6.5 and 6.6 show the modal controllability of the first five modes of the structure as a function of the piezoelectric actuator location on the plate, i.e. the location of one corner of the piezoelectric patch. It can be observed that mode $(1, 1)$ has a maximum modal controllability at

Table 6.2 Ten lowest natural frequencies of the plate

No.	Mode (m,n)	Frequency (Hz)
1	(1,1)	41.9
2	(2,1)	87.1
3	(1,2)	122.4
4	(3,1)	162.4
5	(2,2)	167.6
6	(3,2)	242.9
7	(1,3)	256.5
8	(4,1)	267.9
9	(2,3)	301.7
10	(4,2)	348.3

$x_{11} = 0.36$ m and $y_{11} = 0.26$ m. This corresponds to placing the actuator in the middle of the plate. The result is as expected since the highest strain associated with that mode occurs at that location. The spatial controllability $\mathcal{S}_c$ is plotted in Figure 6.7. It can be observed that an actuator placed in the middle of the plate would provide a considerably high level of spatial controllability. Figure 6.8 shows the spatial controllability for the next five modes for control spillover reduction, $\mathcal{S}_{c2}$. It is desirable to find the location where $\mathcal{S}_{c2}$ is reasonably small, while maintaining a sufficiently high $\mathcal{S}_c$.

The constrained optimization problem is set up as in (6.30):

$$\begin{aligned} &\max_{(x_{11},y_{11})\in\mathcal{R}_1} \quad \mathcal{S}_c(x_{11},y_{11}) \\ &\text{subject to: } \mathcal{M}_{m_i n_i}(x_{11},y_{11}) \geq 50\%,\ i = 1,2,\ldots,5 \\ &\qquad\qquad\quad \mathcal{S}_{c2}(x_{11},y_{11}) \leq 60\%. \end{aligned} \tag{6.52}$$

The minimum level of modal controllability for the first five modes is set at 50% for each mode, while the contributions to the spatial $\mathcal{H}_2$ norm of the next five modes are limited to 60%. Matlab Optimization Toolbox is used to solve the optimization problem, whose optimum solution is found to be $x_{11} = 0.1536$ m and $y_{11} = 0.1418$ m, and associated with that:

$$\begin{aligned} \mathcal{S}_c &= 92.6\% \\ \mathcal{M}_{11} &= 55.1\%; \\ \mathcal{M}_{21} &= 80.6\%; \\ \mathcal{M}_{12} &= 65.3\%; \\ \mathcal{M}_{31} &= 63.4\%; \\ \mathcal{M}_{22} &= 96.1\% \text{ and} \\ \mathcal{S}_{c2} &= 55.6\%. \end{aligned}$$

The above solution shows that the optimal spatial controllability is over

90%, while maintaining the modal controllability of all five modes above 50%. On the other hand, the spatial controllability for control spillover reduction, $\mathcal{S}_{c2}$, is maintained at a level below 60%, which implies that the contributions of those modes to the control spillover will be relatively low. In general, the objective function $\mathcal{S}_c$ has multiple local optima, so a range of initial guesses for x_{11} and y_{11} have been tried to obtain an acceptable result.

Using the same optimization procedure, we can also determine the optimal position and size of each collocated actuator-sensor pair on the plate. In this case, four optimization variables are needed, where two variables are for the position and the other two are for the actuator-sensor size.

6.6.1 *Experiments*

The optimal solution obtained for the placement of the collocated piezoelectric actuator-sensor pair was implemented into a real plate structure. The plate was made from an aluminum alloy, and the structure was held by a support frame. Analytical and finite element analysis using STRAND7 software were performed to ensure sufficient rigidity of the frame up to at least 350 Hz. This would cover the first ten vibration modes of the structure. The plate and frame apparatus were placed on an optical table to minimize the external vibrations affecting the experimental measurements. The experimental apparatus is shown in Figure 6.9. Aluminum shims were used to approximate the pinned (simply-supported) boundary conditions. These shims were inserted along the plate edges as shown in Figure 6.10 and clamped to the frame support that was used to hold the plate in place. The shims allowed some rotational and transverse movement of the plate edges due to their bending. However, it was observed that the transverse movement was small compared to the rotational movement of the edges, so the shims approximated pinned boundary conditions sufficiently well.

A high voltage amplifier, capable of driving highly capacitive loads, was used to supply necessary voltage for the piezoelectric actuator. An HP89410A Dynamic Signal Analyzer and a Polytec PSV-300 Laser Doppler Scanning Vibrometer were used to obtain frequency responses from the plate.

Table 6.3 compares the experimentally measured and the simulated six lowest resonance frequencies of the plate. It can be observed that the difference between the simulation and the experimental results increases at higher frequencies. The error in resonance frequencies of up to several percent was observed from the model. However, the model is reasonably

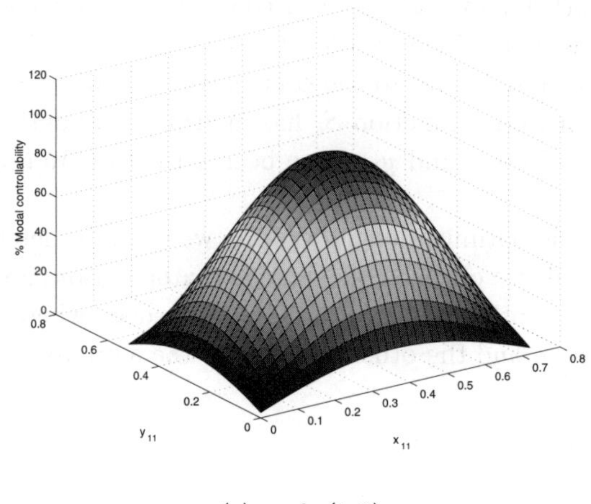

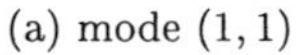
(a) mode $(1, 1)$

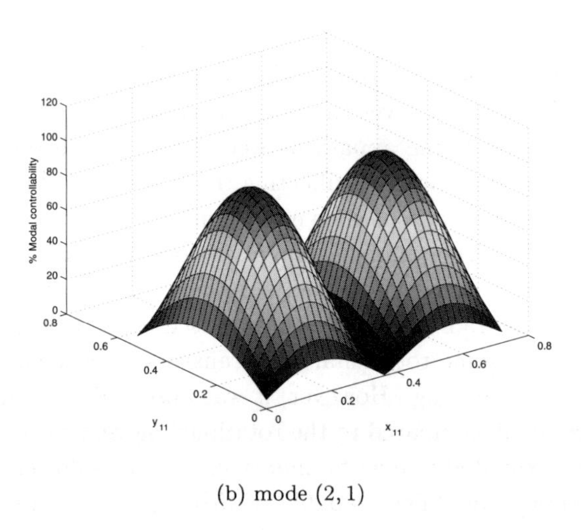

(b) mode $(2, 1)$

Fig. 6.4 Modal controllability: modes 1 and 2

close to the actual plate at low frequencies, considering the experimental results.

Figure 6.11 shows the comparison of the frequency responses of the collocated actuator-sensor system. The dashed line represents the simulation, while the solid line represents the experimental results. The resonance frequencies and damping ratios obtained from the experiments have been used

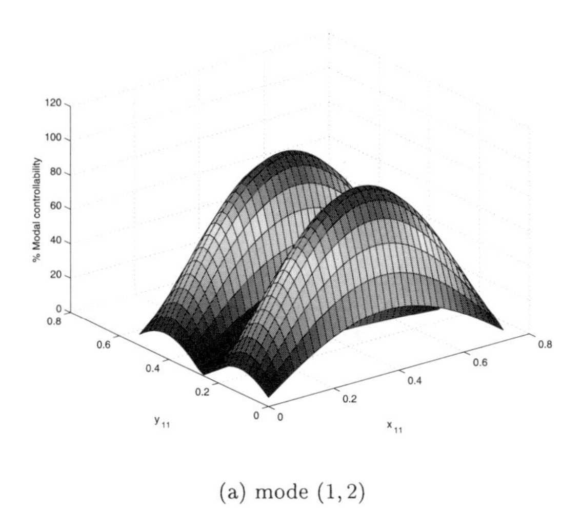

(a) mode $(1, 2)$

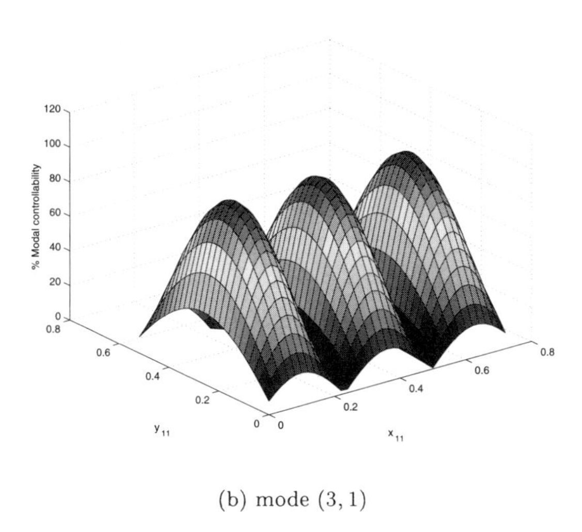

(b) mode $(3, 1)$

Fig. 6.5 Modal controllability: modes 3 and 4

to correct the model. Figure 6.11 demonstrates that the dynamics of the real system are very close to the model.

The effectiveness of the proposed placement methodology can be observed in Figure 6.11. The first five modes (41.8 to 164.3 Hz) show large resonant responses. The result is as expected as the modal controllability/observability of those modes are maintained at reasonably high levels.

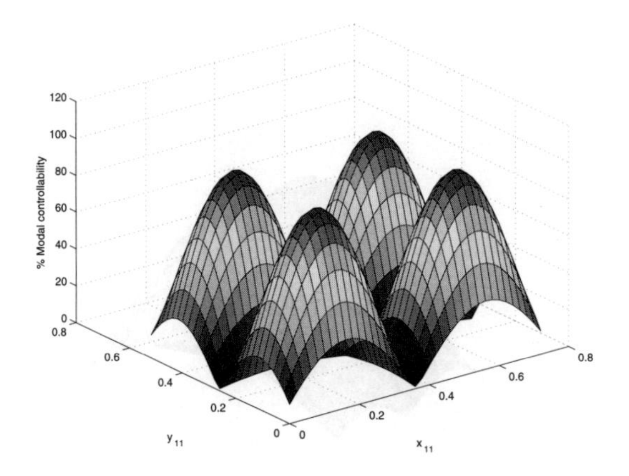

Fig. 6.6 Modal controllability: mode 5 (2,2)

Table 6.3 First six resonance frequencies of the plate

No.	Mode (m,n)	Simulation: ω_{mn} (Hz)	Experiment: ω_{mn} (Hz)	Error (%)
1	$(1,1)$	41.9	41.8	0.2
2	$(2,1)$	87.1	85.9	1.4
3	$(1,2)$	122.4	121.1	1.1
4	$(3,1)$	162.4	159.2	2.0
5	$(2,2)$	167.6	164.3	2.0
6	$(3,2)$	242.9	234.5	3.6

Also, in the optimization, the spatial controllability/observability contribution of the next five modes (234.5 to 327.2 Hz) has been reduced to 55.6% to reduce the spillover effect. It can be observed from the experimental results (Figure 6.11) that the next five modes, with the exception of the sixth mode (at 234.5 Hz), have less resonant responses compared to the resonant responses of the first five modes. Modes 8 (at 256.9 Hz) and 10 (at 327.2 Hz) can hardly be seen in the figure.

The sixth mode at 234.5 Hz has a comparable profile to those of the first five modes. It should be noted that the optimization process only reduces the spatial controllability associated with modes 6 to 10, which can be seen as an average of modal controllability of those modes. As a result, some modes might have relatively high modal controllability levels. To avoid such a problem, one can also try to reduce the modal controllability level corresponding to each of those modes to ensure that each modal controllability is sufficiently low. However, more constraints will result in a more

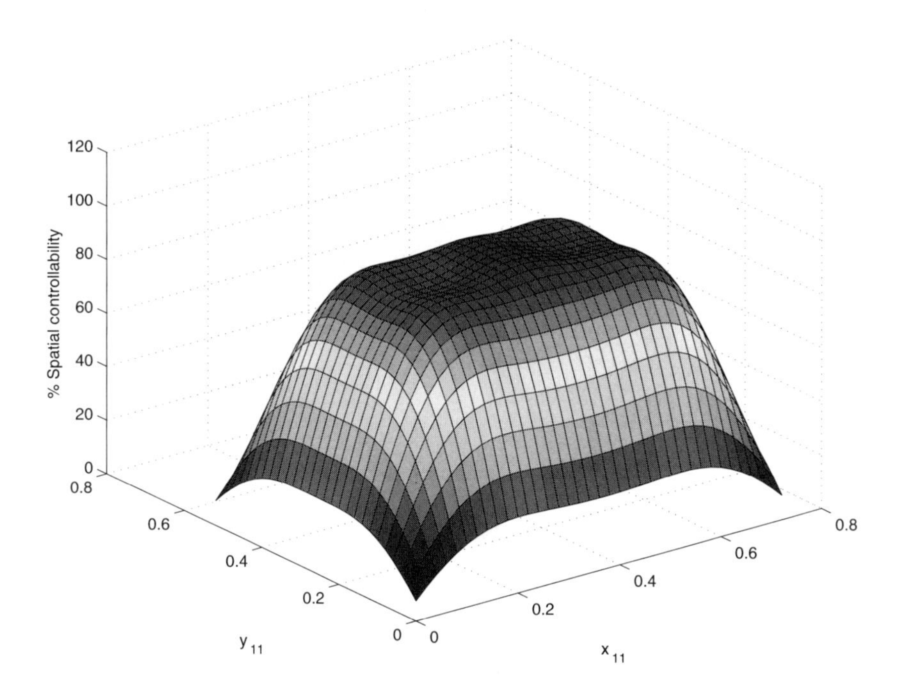

Fig. 6.7 Spatial controllability $\mathcal{S}_c$ - based on the first five modes of the plate

complex optimization problem.

Figure 6.12 shows the mode shapes of the first four vibration modes as obtained from the PSV-300 Laser Doppler Scanning Vibrometer. The mode shapes are reasonably close to the sinusoidal form of mode shapes for a pinned plate, i.e. the plate has sinusoidal shapes in both X and Y directions. The result shows that the use of aluminum shims for the plate pinned boundary conditions is reasonable, especially for low-frequency modes.

6.7 Conclusions

In this chapter we argued that the problem of actuator and sensor placement is important in control of flexible structures. We then introduced and used the notions of spatial controllability and modal controllability for optimal placement of collocated piezoelectric actuator-sensor pairs on a thin flexible plate. The optimization methodology enabled us to place the actuator-sensor pairs at locations that would facilitate the task of designing spatial

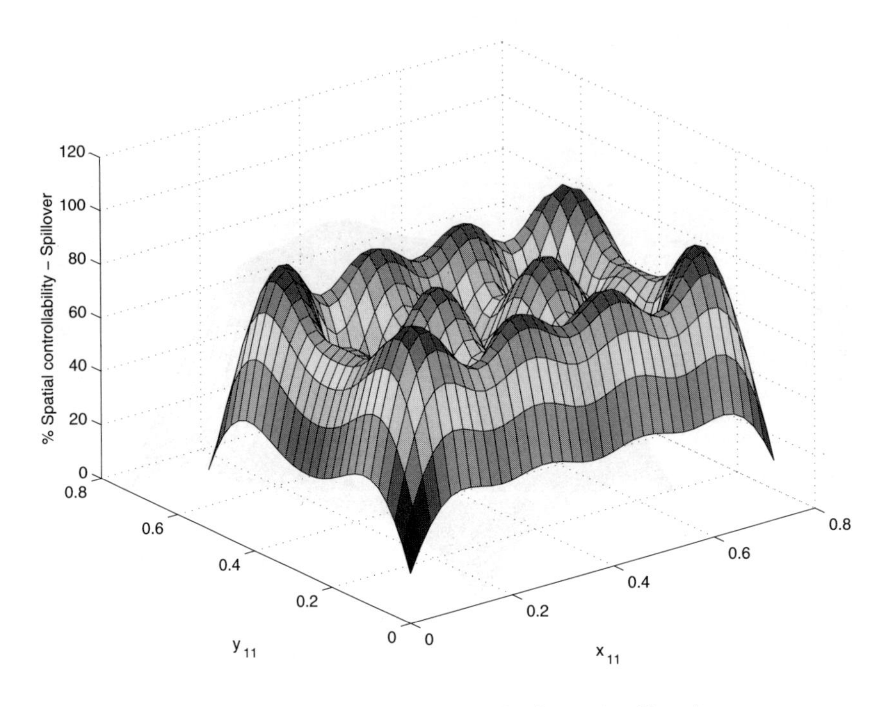

Fig. 6.8 Spatial controllability $\mathcal{S}_{c2}$ (control spillover)

controllers for the system. The proposed methodology was applied to a simply-supported thin plate, and experimental results obtained from the rig were presented and discussed.

Fig. 6.9 Experimental apparatus

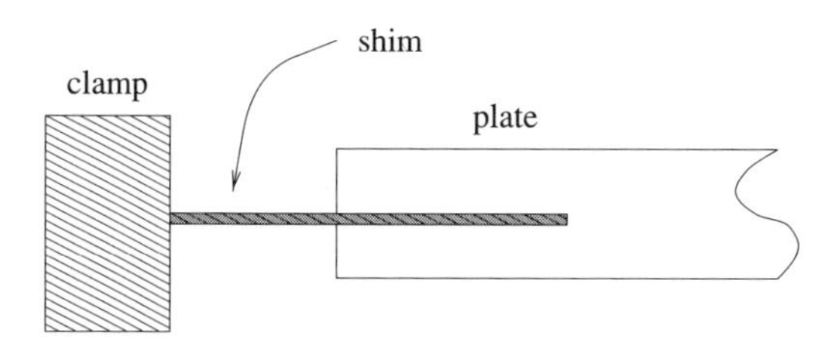

Fig. 6.10 Shims - boundary conditions

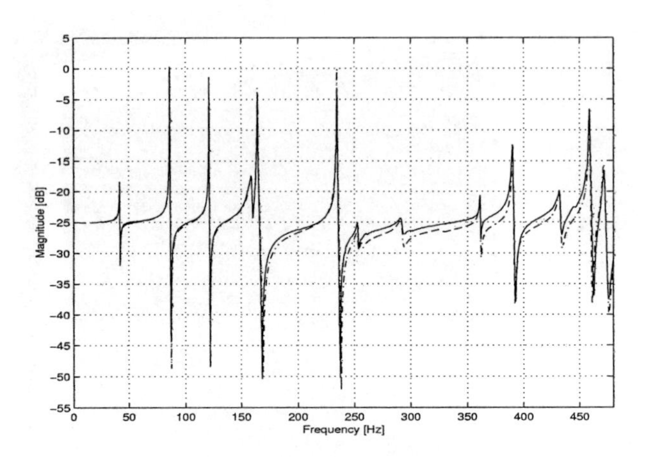

(a) magnitude

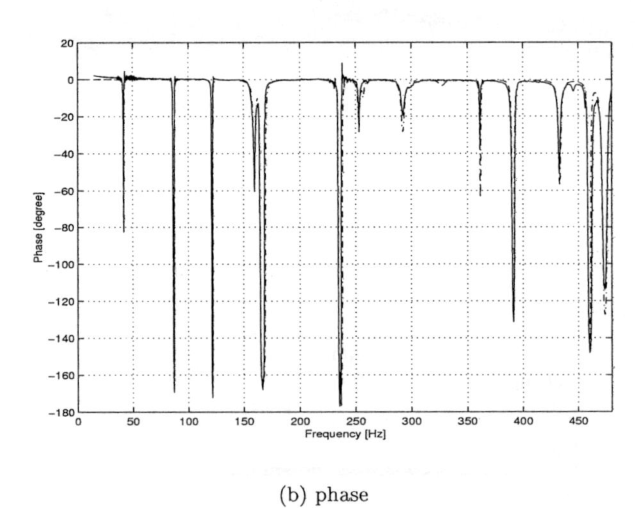

(b) phase

Fig. 6.11 Collocated actuator-sensor system response (voltage - voltage)

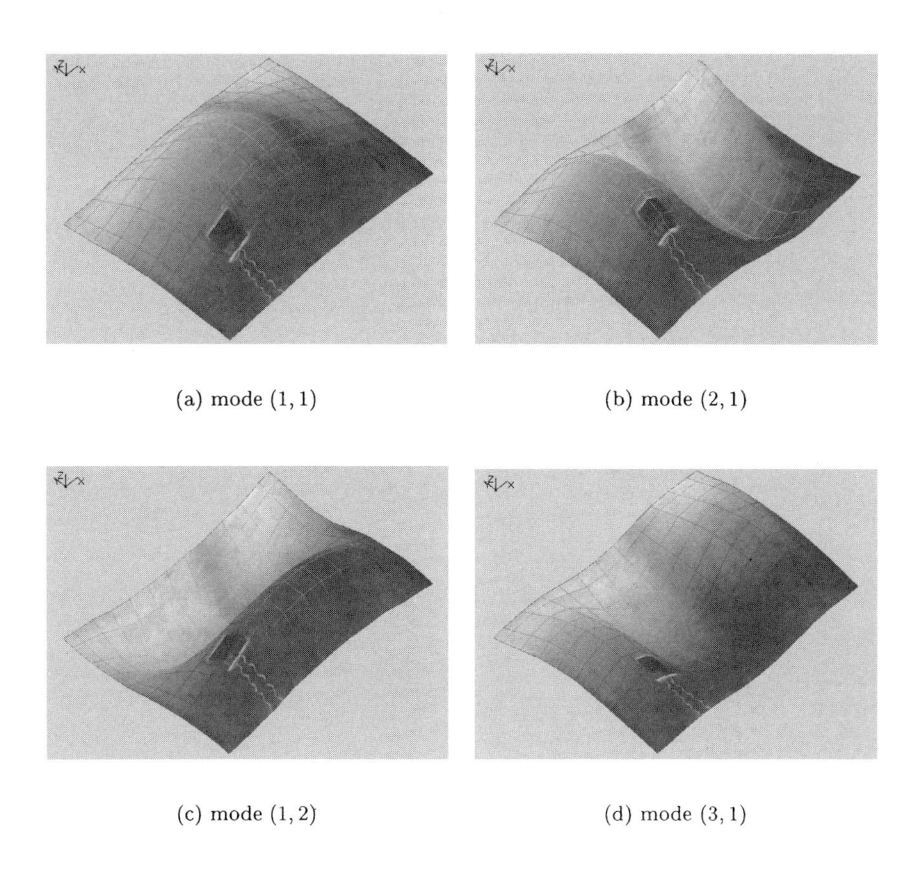

(a) mode $(1,1)$ (b) mode $(2,1)$

(c) mode $(1,2)$ (d) mode $(3,1)$

Fig. 6.12 Mode shapes of the four lowest frequency modes

(c) mode (1, 2) (d) mode (3, 1)

Fig. 6.2 Mode shapes of the four lowest frequency modes

Chapter 7

System Identification for Spatially Distributed Systems

7.1 Introduction

In the analysis and control of distributed parameter systems it is of great benefit to possess a spatial model. That is, a model that describes system dynamics over an entire spatial domain. This chapter is concerned with system identification for the class of distributed parameter systems considered in this book.

The motivation for finding such a model lies in both the fields of analysis and synthesis. During analysis the user may simply wish to observe the mode shapes of the structure, or in a more complete utilization of the model, mathematically estimate the spatial feedback control performance of a system utilizing discrete sensors, actuators, and control objectives. For example, consider Section 5.4 where a standard H_∞ controller [97, 112] is designed to minimize vibration at a single point on a piezoelectric laminate simply supported beam. A spatial model is required to analyze the overall performance of such a controller. The fact that a point-wise controller is shown to provide good local performance but poor spatial performance leads us to the primary application of spatial models, that is, spatial controller synthesis. A number of standard control synthesis variants have emerged that address the control design of spatially distributed systems with discrete sensors and actuators. Recent examples include: spatial feed forward control [69], spatial resonant control [36], spatial H_2 control (Section 5.6) [38], and spatial H_∞ control (Section 5.2) [37].

The modal analysis procedure has been used extensively throughout the literature for obtaining spatial models of structural [67, 23] and acoustic systems [39]. Its major disadvantage is the requirement for detailed physical information regarding the sensors, actuators, and underlying mechanical system. Practical application typically involves the use of experimental data and a non-linear optimization to identify unknown parameters, such as modal amplitudes, resonance frequencies and damping ratios. Even in this case, the descriptive partial differential equations must still be solved (as functions of the unknown parameters) to obtain the mode shapes. This may be difficult or impossible for realistic structural or acoustic systems with complicated boundary conditions.

Another popular technique for obtaining spatial models is that of finite element (FE) analysis [13]. This is an approximate method that results in high order spatially discrete models. If the dynamics of sensors and actuators are known, the integrated model can be cast in a state-space form to facilitate control design and analysis [59]. The approximate nature of finite element modeling eliminates the need for solving descriptive partial differential equations. Detailed information regarding the structures' material properties and boundary conditions is still required. As with the modal analysis procedure, FE models are usually tuned with experimental data [19].

A considerable literature has also developed on the topic of *Experimental Modal Analysis*, (see [62] for a compilation of such methods). These methods can be predominantly described as frequency domain transfer function methods. The system is assumed to consist solely of parallel second order resonant sections. Sensor, actuator, and additional non-modal dynamics are neglected. One of the most popular methods, widely used in commercial frequency domain modal analysis packages, is the rational fraction polynomial method [62]. As a transfer function method, the model is poorly conditioned, incorrectly describes the systems zero dynamics [75], and neglects non-modal dynamics. In addition, all of the mentioned experimental modal analysis techniques neglect the fundamental limitations in spatial sampling, i.e. reconstructed mode shapes can be distorted due to violation of the Nyquist criterion in one and two dimensions.

This chapter introduces an efficient and correct method for identifying the above class of systems directly from measured frequency response data.

7.2 Modeling

The Lagrangian/modal expansion, or Ritz-Kantorovitch method [67], is commonly used to express the spatial deflection of a distributed parameter system as an infinite summation of modes. The modes are a product of two functions, one of the spatial co-ordinate vector $\mathbf{r}$, and another of the temporal t.

$$d\left(t,\mathbf{r}\right)=\sum_{i=1}^{\infty}q_i\left(t\right)\phi_i\left(\mathbf{r}\right), \tag{7.1}$$

where the $q_i(t)$s are the modal displacements, the $\phi_i\left(\mathbf{r}\right)$s are the system eigenfunctions, $d\left(t,\mathbf{r}\right)$ is the displacement at a point, and $\mathbf{r}\in\mathcal{R}$ is a coordinate vector on the spatial domain $\mathcal{R}$. The mode shapes $\phi_i(\mathbf{r})$ must form a complete coordinate basis for the system, satisfy the geometric boundary conditions, and for analytic analysis be differentiable over the spatial domain to at least the degree required by the describing partial differential equations. Many practical systems also obey certain orthogonality conditions. An introduction to the modeling of such systems can be found in Chapter 2.

As discussed in [67] the model (7.1) can also be expressed in the frequency domain as

$$G_y\left(s,\mathbf{r}\right)=\sum_{i=1}^{\infty}\frac{F_i\phi_i\left(\mathbf{r}\right)}{s^2+2\zeta_i\omega_i s+\omega_i^2}, \tag{7.2}$$

where $G_y(s,\mathbf{r})$ is the transfer function from an external force, or for the system considered in this chapter, the applied piezoelectric voltage to the displacement at a point $\mathbf{r}$.

For practical reasons, (7.2) is often truncated to include only a certain number of modes that approximate the response over a limited bandwidth. Chapter 4 introduces a model reduction technique for systems that satisfy certain modal orthogonality conditions. The following truncated model structure is proposed.

$$\widetilde{G}_y\left(s,\mathbf{r}\right)=\sum_{i=1}^{N}\frac{F_i\phi_i\left(\mathbf{r}\right)}{s^2+2\zeta_i\omega_i s+\omega_i^2}+\sum_{i=N+1}^{\infty}k_i\phi_i\left(\mathbf{r}\right), \tag{7.3}$$

where, (referring to Chapter 4), the k_i terms are found by minimizing the spatial $\mathcal{H}_2$ norm of the resulting error system, (ω_c is the retained bandwidth),

$$k_i=\frac{F_i}{2\omega_c\omega_i}\ln\left(\frac{\omega_i+\omega_c}{\omega_i-\omega_c}\right). \tag{7.4}$$

We define the model of a general single input spatially distributed system as

$$\widehat{G}_y(s,\mathbf{r}) = H(s)\left[\sum_{i=1}^{N} \frac{\Phi_i(\mathbf{r})}{s^2 + 2\zeta_i\omega_i s + \omega_i^2} + D(\mathbf{r})\right], \tag{7.5}$$

where $H(s)$ is the concatenation of all non-distributed transfer functions, $\Phi_i(\mathbf{r})$ is the ith mode shape incorporating the modal gain F_i, and $D(\mathbf{r})$ is the feed-through function included to compensate for all higher order truncated contributions to zero dynamics. The filter $H(s)$ is used to model the additional dynamics of sensors, actuators, and for example, anti-aliasing filters. In this work $H(s)$ is not identified automatically.

The objective will be to identify the parameters $\theta = \begin{bmatrix} \Phi_i(\mathbf{r}) & D(\mathbf{r}) & \zeta_i & \omega_i \end{bmatrix}$ from a number of measured spatially distributed point-wise frequency responses,

$$G_y(j\omega,\mathbf{r}) \quad \begin{matrix} \mathbf{r} \in \{\mathbf{r}_1,\ldots,\mathbf{r}_{N_r}\} \in \mathcal{R} \\ \omega \in \{\omega_1,\ldots,\omega_{N_\omega}\}, \end{matrix} \tag{7.6}$$

where N_r is the number of measured spatial locations and N_ω is the number of measured frequency points per location.

The system (7.5) has a corresponding state space representation:

$$\begin{aligned} \dot{\mathbf{x}}(t) &= \mathbf{A}\mathbf{x}(t) + \mathbf{B}u(t) \\ d(t,\mathbf{r}) &= \mathbf{C}(\mathbf{r})\mathbf{x}(t) + D(\mathbf{r})u(t), \end{aligned} \tag{7.7}$$

where $\mathbf{C}(\mathbf{r}) = \begin{bmatrix} \Phi_i(\mathbf{r}) \; 0 \cdots \Phi_N(\mathbf{r}) \; 0 \end{bmatrix}$, $\mathbf{B} = \begin{bmatrix} 0 \; 1 \cdots 0 \; 1 \end{bmatrix}^T$, $D(\mathbf{r})$ is a scalar function of $\mathbf{r}$, N is the number of modes to be identified, and

$$\mathbf{A} = \begin{bmatrix} 0 & 1 & & 0 & 0 \\ -\omega_1^2 & -2\zeta_1\omega_1 & & 0 & 0 \\ & & \ddots & & \\ 0 & 0 & & 0 & 1 \\ 0 & 0 & & -\omega_N^2 & -2\zeta_N\omega_N \end{bmatrix} \in \mathbf{R}^{2N\times 2N}. \tag{7.8}$$

7.3 Spatial sampling

Considering the model structure (7.5), the spatial functions $\Phi_i(\mathbf{r})$ and $D(\mathbf{r})$ must be reconstructed from their identified samples. For a uniformly sampled one-dimensional system, the samples of our continuous functions $\Phi_i(\mathbf{r})$

and $D(\mathbf{r})$ are

$$\begin{array}{ll} \Phi_i(r) & r = n\,\Delta r \in \mathcal{R} \\ D(r) & n \in \{0, 1, ..., N_r\}, \end{array} \tag{7.9}$$

where the scalar r specifically denotes a one-dimensional system and Δr is the spatial sampling interval.

There are a number of options available for reconstructing the continuous functions, two of which are traditional linear reconstruction and spline reconstruction. The following two subsections, 7.3.1 and 7.3.2, examine the application of each technique to the two cases of band-limited and non-band-limited functions. The aim is to quantify the expected mean square difference between the original continuous function and its corresponding reconstruction. This will allow us to evaluate the required spatial sampling interval as a function of the permissible error. An example of this procedure is performed for a simply supported beam in Section 7.3.3.

7.3.1 *Whittaker-Shannon reconstruction*

The discrete magnitude spectra of a band-limited spatial function $f(r)$ is shown in Figure 7.1. To satisfy the Nyquist sampling criterion, the spatial sampling frequency $\frac{2\pi}{\Delta r}$ (in $\frac{rad}{m}$) must be greater than twice the highest frequency component of $f(r)$ [47]. Shannon's Reconstruction Theorem states that $f(r)$ can be reconstructed from its samples,

$$f(r) = \Delta r \sum_{k=-\infty}^{\infty} f(k\Delta r) \frac{\sin\left(\frac{\pi}{\Delta r}(r - k\Delta r)\right)}{\pi(r - k\Delta r)}. \tag{7.10}$$

Theoretically, a perfect reconstruction is possible, however, in practice, there are two significant sources of degradation:

(1) For finite time signals, truncating the summation (7.10) introduces a systematic *truncation error*. Expressions for bounding the truncation error and references to relevant work can be found in [47].
(2) In many practical situations the samples will also contain an additive stochastic disturbance. An expression for the mean integral squared reconstruction error (MISE) experienced when recovering a signal from its corrupted samples can be found in [86]. It is also shown that Shannon reconstruction is not a consistent estimator for band-limited signals recovered from noisy samples, i.e. as the number of signal samples approaches ∞, the MISE does not approach zero, in fact the error diverges

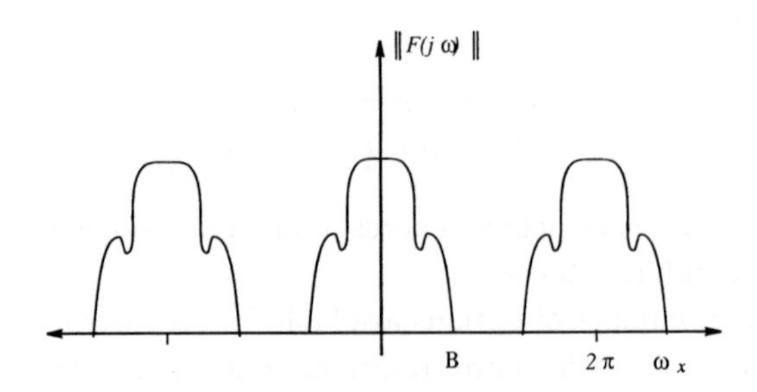

Fig. 7.1 Discrete magnitude spectra of an over sampled band-limited function

and also approaches ∞. Convergent estimators for such scenarios can be found in [87] and [50].

In general, the spatial function $f(r)$ will not be band-limited. Referring to Section 2.6.2 and Chapter 4 respectively, examples include, the mode shapes of a cantilever beam, and the feed-through function for a simply supported beam [75]. Since the samples are obtained indirectly from point-wise frequency response data, no form of low-pass filtering is possible. The objective of the following will be to quantify the under-sampling error as a function of the spatial sampling interval. In their paper reviewing sources of error in linear reconstruction, Thomas and Liu [102] present an expression for the mean square reconstruction error as a function of the power spectral density outside the Nyquist range. The following expression assumes the absence of the optimal low-pass filter, which in our application, cannot be applied to the continuous signal.

$$||f(r) - Q^s\ f(r)||_2 = \left[\frac{1}{\pi}\int_{|\omega_r|>\frac{\pi}{\Delta r}} |F(j\omega_r)|^2\ d\omega_r\right]^{\frac{1}{2}}, \tag{7.11}$$

where $Q^s f(r)$ is the Shannon representation of the sampled function, ω_r is the spatial frequency in radians per meter and $|F(j\omega_r)|^2$ is the power spectral density of $f(r)$. In the case where the optimal pre-filter can be applied, the RHS* of equation 7.11 is reduced by one half.

*The part of an equation on the right hand side of the equals sign.

7.3.2 Spline reconstruction

In recent years, splines have been recognized for their usefulness in curve and surface fitting problems [53, 107]. A function $f(r)$ can be approximately reconstructed from a spline basis $\varphi(r)$, with coefficients $c(k)$ derived from $f(k\Delta r)$.

$$\overset{sp}{Q^n} f(r) = \sum_{k\in Z} c(k)\varphi^n(\frac{r}{\Delta r} - k), \tag{7.12}$$

where $c(k) \in l_2$ are the (finite square summable) spline coefficients, $\overset{sp}{Q^n} f(r)$ is the spline reconstruction of $f(r)$ and $\varphi^n(r)$ is the spline generating function. We will limit our choice of generating functions to the nth degree β-splines (of order $n+1$) [107]. The condition $c(k) \in l_2$ ensures that $\overset{sp}{Q^n} f(r)$ is a well-defined subspace of L_2, the set of square integrable functions, a considerably larger space than the traditional Shannon space of band-limited functions. References [108] and [42] present a unified sampling theory for a wide class of approximation operators. In likeness to the Shannon sampling theorem, the optimal spline reconstruction involves an optimal pre-filtering of the continuous signal before sampling and reconstruction by the chosen spline basis. The results in this area, including expressions for the root mean square (RMS) error, are summarized in [107]. The technique of quantitative Fourier analysis can be applied to quantify the RMS reconstruction error [6]. The sampling phase averaged error is given by,

$$\left\| f(r) - \overset{sp}{Q^n} f(r) \right\|_2 = \left[\frac{1}{2\pi} \int_{-\infty}^{\infty} |F(j\omega_r)|^2 \, E^n(\Delta r \; \omega_r) \, d\omega_r \right]^{\frac{1}{2}}, \tag{7.13}$$

where $E^n(\Delta r \; \omega_r)$ is defined as the frequency error kernel and is a function of the interpolant and Δr. Analytic expressions for $E^n(\Delta r \; \omega_r)$ have been given for the β-splines of order up to 6 [6].

In our application where there is no access to the continuous signal, we cannot apply the optimal pre-filter nor achieve the optimal (least squares) fit by projecting our signal onto the approximation space [107]. Instead, we shall simply perform an interpolation. The penalty in doing so is illustrated in Figure 7.2, where the error kernels $E^n(\omega_r)$ for spline and Shannon reconstruction, optimal and interpolation, are shown for $\Delta r = 1$. It can be observed that although the spline interpolant error is globally greater than that of the projector, within the Nyquist range $|\omega_r| < \pi$ the difference is slight. In analogy to Shannon reconstruction, for frequencies beyond the Nyquist rate, the magnitude of the spline interpolant error kernel approaches twice that of the projector.

The spline basis functions also have some interesting variational properties. It is well known that interpolation by the Shannon basis in the presence of effects such as truncation or additive high-frequency sample noise, tends to result in an overly 'peaky' or oscillatory reconstruction. In contrast, spline interpolation (in a certain sense [95, 107]) is the interpolant that oscillates the least. The cubic spline is a special case, since it minimizes the 2-norm of error's second derivative and it possesses the property of minimum curvature [107]. As this property is also shared by constrained thin elastic beams and plates, it is natural to reason that cubic splines may be well suited to approximating mechanical functions, such as the mode shapes of a simply-supported beam.

In the case of noisy samples, we can achieve some degree of immunity by relaxing the interpolation condition and imposing a smoothness constraint, i.e. for the cubic splines, by minimizing

$$\sum_k \left(f(k\Delta r) - \overset{sp}{Q^n} f(k\Delta r) \right)^2 + \lambda \int_0^L \left(\frac{d^2 \overset{sp}{Q^n} f(r)}{dr^2} \right)^2, \tag{7.14}$$

where the second term is a measure of the smoothness. The parameter λ is based on the additive noise variance [107].

7.3.3 *Spatial sampling of a simply-supported beam*

This chapter demonstrates how the results presented in Section 7.3.2 can be applied to spatial systems. We present an example analysis for the simply-supported beam described in Section 4.8. The objective is to arrive at a point where Equations (7.11) and (7.13) can be applied. Both expressions require only the function's power spectral density.

7.3.3.1 *Mode shapes*

The mode shapes of a simply-supported beam are given by [67]

$$\phi_i(r) = \sqrt{\frac{2}{\rho A_r L}} \sin\left(\frac{i\pi r}{L}\right) \tag{7.15}$$

$$= \alpha \sin\left(\frac{i\pi r}{L}\right), \tag{7.16}$$

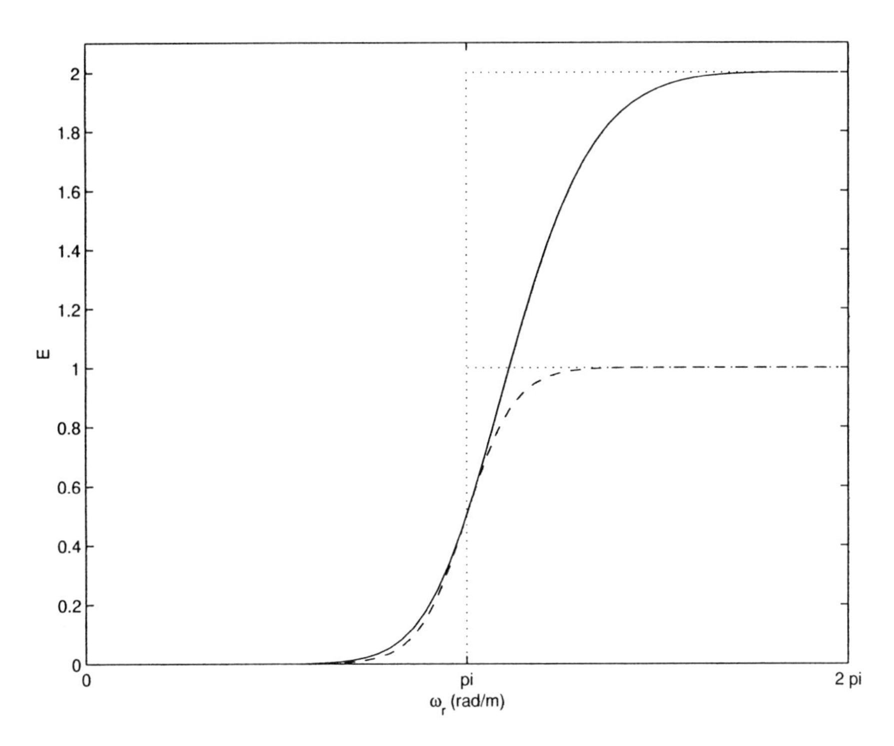

Fig. 7.2 Unit sampled, cubic spline error kernels. Optimal (with pre-filter) (- -), interpolation (–). Shannon reconstruction ($\cdots$) $E(|\omega| > \pi) = 1$, interpolation ($\cdots$) $E(|\omega| > \pi) = 2$

where ρ is the material density, A_r is the cross-sectional area and L is the length of the beam. The spatial spectra of $\sum_{i=1}^{N} \phi_i(r)$ is impulsive and can be easily determined,

$$\mathcal{F}\left\{\sum_{i=1}^{N} \phi_i(r)\right\} = j\pi\sqrt{\frac{2}{\rho A_r L}} \sum_{i=1}^{N} \left[\delta(\omega_r + \frac{i\pi}{L}) - \delta(\omega_r - \frac{i\pi}{L})\right]. \qquad (7.17)$$

The highest frequency component of $\phi_i(r)$, $i \in \{1, \ldots, N\}$ is $\frac{N\pi}{L}$. Thus, if we were to apply Shannon's Theorem† to reconstruct N mode shapes of a simply-supported beam,

$$\frac{2\pi}{\Delta r} > 2\frac{N\pi}{L}, \qquad (7.18)$$

†Neglecting truncation errors.

that is,

$$\Delta r < \frac{L}{N}. \tag{7.19}$$

This simple and complete result applies in general to a sub-class of the systems (7.2). Such systems are characterized by sinusoidal mode shapes. Examples include uniform beams and strings in one dimension, plates in two dimensions and closed acoustic systems in three dimensions.

7.3.3.2 *The feed-through function $D(r)$*

The feed-through function $D(r)$ can be found analytically for systems of the form (7.2).

$$D(r) = \sum_{i=N+1}^{\infty} k_i \phi_i(r), \tag{7.20}$$

where $\phi_i(r)$ is given by (7.15) and k_i is given by (7.4). We can think of (7.20) as being equivalent to the Fourier series,

$$D(r) = \sum_{i=-\infty}^{\infty} c_i e^{\frac{j2\pi ir}{T_r}}, \tag{7.21}$$

where $T_r = 2L$ is the period of repetition,

$$c_i = \begin{cases} \frac{j}{2}\alpha \frac{F_i}{2\omega_c \omega_i} \ln\left(\frac{\omega_i + \omega_c}{\omega_i - \omega_c}\right) & i \in \{\ldots, -N-2, -N-1\} \\ 0 & i \in \{-N, \ldots, N\} \\ \frac{-j}{2}\alpha \frac{F_i}{2\omega_c \omega_i} \ln\left(\frac{\omega_i + \omega_c}{\omega_i - \omega_c}\right) & i \in \{N+1, N+2, \ldots\}. \end{cases} \tag{7.22}$$

The complex coefficients c_i reveal the spatial Fourier transform of $D(r)$.

$$\mathcal{F}\{D(r)\} = \mathcal{F}\left\{\sum_{i=N+1}^{\infty} k_i \phi_i(r)\right\} \tag{7.23}$$

$$= \sum_{i=-\infty}^{\infty} 2\pi c_i \delta(\omega_r - i\omega_f), \tag{7.24}$$

where

$$\omega_f = \frac{2\pi}{T_r} = \frac{\pi}{L}. \tag{7.25}$$

That is,

$$\mathcal{F}\{D(r)\} = \sum_{i=-\infty}^{\infty} 2\pi c_i \delta(\omega_r - i\frac{\pi}{L}). \tag{7.26}$$

Immediately, by the properties of the Fourier transform, we learn some characteristics of the feed-through function $D(r)$.

(1) As verification, $\mathcal{F}\{D(r)\} = d(j\omega_r) = d(-j\omega_r)^* \Leftrightarrow \text{Im}\{D(r)\} = 0$, which is known *a priori*. Also $\text{Re}\{d(j\omega_r)\} = 0$ and $|d(j\omega_r)|$ are even functions of ω_r [92] and $\text{Im}\{d(j\omega_r)\}$ is an odd function of ω_r [92].
(2) Since $\mathcal{F}\{D(r)\}$ is purely imaginary, $D(r)$ is an odd function. It is true in general that $\text{Re}\{d(j\omega_r)\} = 0 \Leftrightarrow D(r) = -D(-r)$ [92].
(3) $D(r)$ is periodic with period $2L$.

As $\mathcal{F}\{D(r)\}$ does not have compact support on the interval $(-j\infty, j\infty)$, $D(r)$ cannot be exactly reconstructed with any finite number of samples. It is also obvious from (7.26) that the spectra of $D(r)$ lies completely outside the bandwidth of the mode shapes, thus dictating the spatial sampling requirements of the system.

We can now apply Equation (7.13) to determine the required spatial sampling interval. For a periodic signal $g(r)$, the energy density per unit frequency is given by [22]

$$|G(f)|^2 = T \sum_{n \in Z} |c_n|^2 \delta(f - n\frac{1}{T}), \tag{7.27}$$

where T is the period, $G(f)$ denotes the Fourier transform and c_n are the Fourier coefficients of $g(r)$. By making a change of variables we can find the power spectral density of $D(r)$,

$$|\mathcal{F}\{D(r)\}|^2 = 2\pi T_r \sum_{i \in Z} |c_i|^2 \delta(\omega_r - i\frac{2\pi}{T_r}). \tag{7.28}$$

Hence, from equation (7.13), the error in reconstructing $D(r)$ from an nth order spline basis can be obtained,

$$\left\| D(r) - \overset{sp}{Q^n}\ D(r) \right\|_2 = \left[2L \int_{-\infty}^{\infty} \left(\sum_{i \in Z} |c_i|^2 \delta(\omega_r - i\frac{\pi}{L}) \right) E^n(\Delta r\ \omega_r)\ d\omega_r \right]^{\frac{1}{2}} \tag{7.29}$$

$$= \left[2L \sum_{i \in Z} |c_i|^2\ E^n(i\pi\frac{\Delta r}{L}) \right]^{\frac{1}{2}}, \tag{7.30}$$

where $\overset{sp}{Q^n} D(r)$ is the spline reconstruction of $D(r)$. The error kernel for a cubic spline $E^3(i\pi\frac{\Delta r}{L})$ is plotted together with the equivalent Shannon kernel in Figure (7.2).

We can also apply Parseval's equality to find the mean square value of $D(r)$ over one period

$$\sum_{i=-\infty}^{\infty} |c_i|^2 = \frac{1}{2L}\int_{-L}^{L} |D(r)|^2 \, dr. \tag{7.31}$$

We now consider a specific example: the simply-supported beam described in Section 4.8, where 3 modes are retained for identification. The feed-through function resulting from an analytic model derived in Section 4.8 is shown in Figure 7.3. The RMS value of the reconstruction error (L_2 norm on $[-L, L]$) is plotted against the sampling interval Δr in Figure 7.4. As the sampling interval increases, the RMS error approaches the RMS value of the continuous function‡. This plot can be used to select a spatial sampling interval that achieves some error specification on $D(r)$.

7.3.3.3 *Other considerations*

The above analysis has considered only a one-dimensional system. The Shannon sampling theorem is easily extended to multi-variate functions [47]. By using tensor-product basis functions, spline sampling theory is extended in a similar fashion [107]. Both techniques require an equidistant sampling grid and are based on the application of uni-variate results in each dimension. For irregular sampling and other complicated reconstructions (e.g. by blending functions [53], or finite element methods [53]) no such results are known.

In the previous subsection, i.e. Section 7.3.3, the sampling limitations for a simply supported beam have been derived. Even when the mode shapes are known *a priori*, this analysis can be difficult to perform. For the practitioner, we offer a rough rule of thumb.

(1) Estimate, by means of a similar system or finite element analysis, the highest significant spatial frequency component of the highest order mode to be identified.
(2) Consider the feed-through function $D(\mathbf{r})$. Assume that its highest significant frequency component is three times that estimated in step (1).

‡In this analysis we have considered $D(r) \notin L_2$. This arises from the periodic nature of the mode shapes. When we refer to the RMS or mean square value of such signals, we are implicitly referring to the RMS or mean square value over a single period.

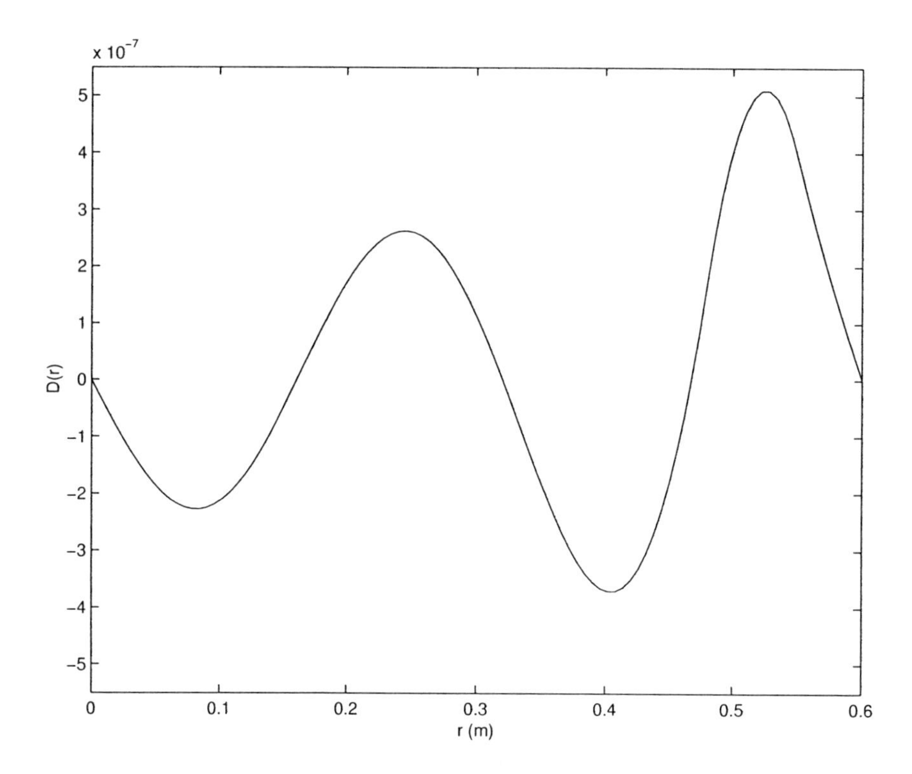

Fig. 7.3 Analytic feed-through function for the beam described in Section 4.8

(This step is suggested on the experience of studying and identifying a number of such systems).

(3) Sample the structure as would be done in practice for a function with spatial bandwidth derived in step (2). Taking into consideration the limited domain of the structure, (allowing for truncation errors), this would normally be between 2 to 5 times the rate suggested by the Nyquist criterion.

7.4 Identifying the system matrix

The first step in the identification procedure is to obtain an estimate for $\mathbf{A}$, the system matrix whose eigenvalues reveal the parallel dynamics of each mode. On first inspection, this problem may appear trivial as the transfer function obtained from a single frequency response would perform the task.

For spatially distributed systems we need to redefine our measures

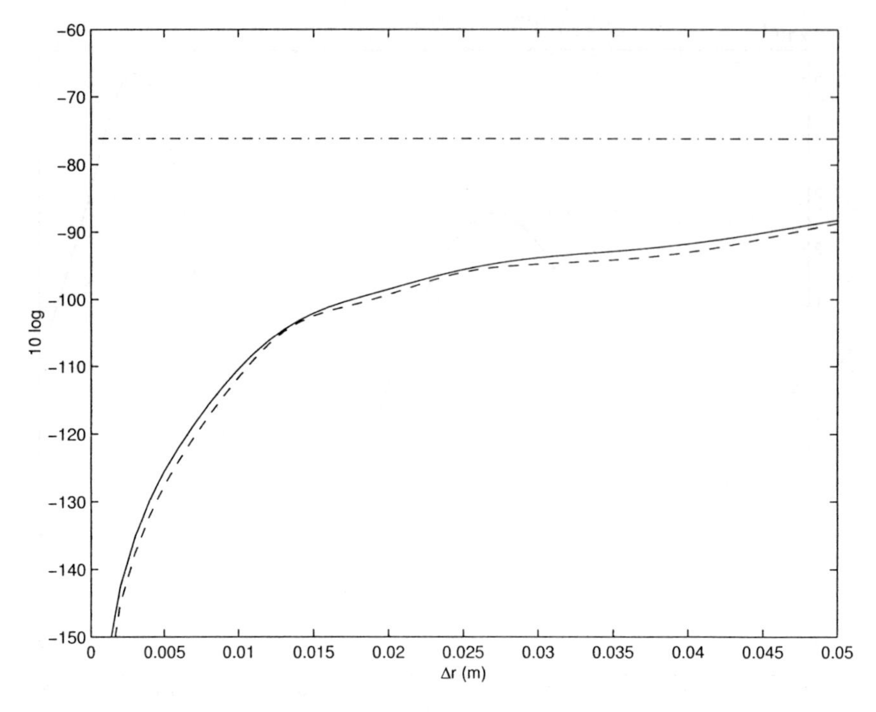

Fig. 7.4 The RMS reconstruction error $||D(r) - Q^n\ D(r)||_2$ plotted against the spatial sampling interval Δr. The dashed-dot line indicates the RMS value of the function $D(r)$

of model quality and stochastic performance. In essence, the two main sources of error in the identification arise from measurement noise and slight changes in system dynamics over the spatial domain. Intuitively, we would like to distribute the resulting model error in a similar, equally distributed fashion. The task of quantifying such errors is the subject of current research.

The problem can be cast as a MIMO system identification problem where each point is regarded as a single output. In the case of a two-dimensional system, where a large number of point-wise frequency response measurements are available, it may be necessary to limit the data space by selecting only a subset of the available points. The *virtual system* as seen by the system identification algorithm has a single input and $\widetilde{N}_r$ outputs, where $\widetilde{N}_r$ may be equal to N_r or less than N_r if the data set is to be truncated. The frequency response of such a system is similar to (7.6) and can be expressed as

$$G_y(j\omega,\mathbf{r}) \quad \begin{array}{l} \mathbf{r} \subset \{\mathbf{r}_1,\ldots,\mathbf{r}_{N_r}\} \in \mathcal{R} \\ \omega \in \{\omega_1,\ldots,\omega_{N_\omega}\}. \end{array} \tag{7.32}$$

For generality, we treat the identification algorithm as a general matrix function of the data, i.e. $\mathbf{A} = f\left(G_y\left(j\omega,\mathbf{r}\right)\right)$.

Methods that identify state space models by exploiting geometric properties of the input and output sequences are commonly known as subspace methods. These methods have received considerable attention in the literature, (see [109] for a survey of time domain methods). The reader is referred to [60] and [65] for a full discussion of frequency domain techniques. Frequency domain subspace-based algorithms have proven particularly useful for identifying high order multi-variable resonant systems [64]. In this chapter the algorithm introduced in [65] and summarized in Appendix 7.7 will be employed.

7.5 Identifying the mode shapes and feed-through function

Samples of the spatial modal and feed-through functions are first identified from the frequency response data and system matrix. The continuous functions are then approximated by linear or spline reconstruction.

7.5.1 *Identifying the samples*

Samples of the spatial functions will now be identified from the available frequency response data. First, in order to simplify notation, we make a number of definitions.

The spatial response matrix is defined as

$$\mathbf{G} = \begin{bmatrix} G_y(j\omega_1,\mathbf{r}_1) & \cdots & G_y(j\omega_1,\mathbf{r}_{N_r}) \\ \vdots & \ddots & \vdots \\ G_y(j\omega_{N_\omega},\mathbf{r}_1) & \cdots & G_y(j\omega_{N_\omega},\mathbf{r}_{N_r}) \end{bmatrix} \in \mathbf{C}^{N_\omega \times N_r}. \tag{7.33}$$

The dynamic response matrix is defined as

$$\mathbf{P}^{tf} = \begin{bmatrix} P_1^{-1}(j\omega_1) & \cdots & P_N^{-1}(j\omega_1) \\ \vdots & \ddots & \vdots \\ P_1^{-1}(j\omega_{N_\omega}) & \cdots & P_N^{-1}(j\omega_{N_\omega}) \end{bmatrix} \in \mathbf{C}^{N_\omega \times N}, \tag{7.34}$$

where $P_i^{-1}(j\omega)$ is the response of the ordered i^{th} mode dynamics found from the system matrix $\mathbf{A}$

$$P_i^{-1}(j\omega) = \left.\frac{1}{[s + (\alpha_i + j\sigma_i)]\,[s + (\alpha_i - j\sigma_i)]}\right|_{s=j\omega}. \tag{7.35}$$

The modal function matrix is defined as

$$\Psi = \begin{bmatrix} \Phi_1(\mathbf{r}_1) & \cdots & \Phi_1(\mathbf{r}_{N_r}) \\ \vdots & \ddots & \vdots \\ \Phi_N(\mathbf{r}_1) & \cdots & \Phi_N(\mathbf{r}_{N_r}) \end{bmatrix} \in \mathbf{R}^{N \times N_r}. \tag{7.36}$$

The feed-through vector is defined as

$$\mathbf{D} = \begin{bmatrix} D(\mathbf{r}_1) & \cdots & D(\mathbf{r}_{N_r}) \end{bmatrix} \in \mathbf{R}^{1 \times N_r}. \tag{7.37}$$

We can form the following complex matrix equation

$$\mathbf{G} = \begin{bmatrix} \mathbf{P}^{tf} & 1_{N_\omega \times 1} \end{bmatrix} \begin{bmatrix} \widehat{\Psi} \\ \widehat{\mathbf{D}} \end{bmatrix}, \tag{7.38}$$

where $1_{N_\omega \times 1}$ denotes a matrix with dimension $N_\omega \times 1$, whose entries are all 1. Equation (7.38) has a unique least squares solution if $N_\omega \geqslant N$, this condition is automatically satisfied if the restrictions for the subspace estimation in Section 7.4 are met, i.e., if $N_\omega \geqslant q + p$, where p is the model order and q is the auxiliary order [65]. Since we are interested in real valued functions we restrict the matrices $\widehat{\Psi}$ and $\widehat{D}$ accordingly.

7.5.2 *Linear reconstruction*

Here the ordering and dimension of the co-ordinate vector $\mathbf{r}$ becomes important. For notational simplicity, we assume $\mathbf{r}$ is single dimensional. Shannon's formula for linear reconstruction can be restated in context.

$$\begin{aligned} \Phi_i(r) &= \Delta r \sum_{k=0}^{N_x} \Phi_i(r_k) \frac{\sin\left(\frac{\pi}{\Delta r}(r - r_k)\right)}{\pi(r - r_k)} \\ &= [\Phi_i(r_1) \ \cdots \ \Phi_i(r_{N_r})] \begin{bmatrix} \text{sinc}\ \left(\frac{\pi}{\Delta r}(r - r_1)\right) \\ \vdots \\ \text{sinc}\ \left(\frac{\pi}{\Delta r}(r - r_{N_r})\right) \end{bmatrix}. \end{aligned} \tag{7.39}$$

$D(r)$ can be reconstructed in a similar fashion. For convenience we write an equation describing all spatial functions

$$\begin{bmatrix} \Phi_i(r) \\ \vdots \\ \Phi_N(r) \\ D(r) \end{bmatrix} = \begin{bmatrix} \widehat{\Psi} \\ \widehat{\mathbf{D}} \end{bmatrix} \begin{bmatrix} \text{sinc } \left(\frac{\pi}{\Delta r}(r - r_1)\right) \\ \vdots \\ \text{sinc } \left(\frac{\pi}{\Delta r}(r - r_{N_r})\right) \end{bmatrix} = \begin{bmatrix} \widehat{\Psi} \\ \widehat{\mathbf{D}} \end{bmatrix} \mathbf{B}_r(r), \tag{7.40}$$

where $\mathbf{B}_r(r)$ is the basis of reconstruction.

The spatial system can be written in state space form as

$$\begin{aligned} \dot{\mathbf{x}} &= \mathbf{A}\mathbf{x} + \mathbf{B}u \\ Y(r) &= \mathbf{B}_r(r)'\widehat{\Psi}'J\mathbf{x} + \widehat{\mathbf{D}}\mathbf{B}_r(r)u, \end{aligned} \tag{7.41}$$

where $J = \begin{bmatrix} e_1' \; e_3' \; \cdots \; e_{(2N-1)}' \end{bmatrix}' \in \mathbf{R}^{N \times 2N}$ and e_i is the i shifted unit impulse, e.g., $e_3 = \begin{bmatrix} 0 \; 0 \; 1 \; 0 \; \cdots \; 0 \end{bmatrix}$. Note the equivalence of system (7.41) to (7.7), where $\mathbf{B}_r(r)'\widehat{\Psi}'J$ and $\widehat{\mathbf{D}}\mathbf{B}_r(r)$ represent the identified function matrix $\mathbf{C}(r)$ and feed-through function $\mathbf{D}(r)$.

7.5.3 *Spline reconstruction*

The spline reconstructed system is similar to (7.41) with the exception that the function samples $\begin{bmatrix} \widehat{\Psi} \\ \widehat{\mathbf{D}} \end{bmatrix}$ and reconstruction basis $\mathbf{B}_r$ are replaced by the spline coefficients and chosen spline basis.

7.5.3.1 *Finding the spline coefficients $c(k)$*

Many standard procedures exist for finding the spline coefficients $c(k)$ as defined in Equation (7.12). The reader is referred to reference [107] for an overview of such techniques.

7.5.3.2 *Summary*

After computing the spline coefficients for each mode, the spatial system can be expressed in state space form

$$\begin{aligned}\dot{\mathbf{x}} &= \mathbf{A}\mathbf{x} + \mathbf{B}u \\ Y(r) &= B^n(r)'\mathbf{C}_s'J\mathbf{x} + \mathbf{D}_s B^n(r)u,\end{aligned} \tag{7.42}$$

where $\mathbf{D}_s = \begin{bmatrix} c_d(1) \; c_d(2) \; c_d(N_r) \end{bmatrix}$ the spline coefficients of the feed-through function $D(r)$, $\mathbf{C}_s$ is the matrix containing the spline coefficients for each mode and $B^n(r)$ is the spline reconstruction basis.

$$\mathbf{C}_s = \begin{bmatrix} c_1(1) & \cdots & c_1(N_r) \\ \vdots & \ddots & \vdots \\ c_N(1) & \cdots & c_N(N_r) \end{bmatrix} \tag{7.43}$$

$$B^n(r) = \begin{bmatrix} \beta^n(\frac{r}{\Delta r}) \\ \beta^n(\frac{r}{\Delta r} - 1) \\ \vdots \\ \beta^n(\frac{r}{\Delta r} - (N_r - 1)) \end{bmatrix} \tag{7.44}$$

7.6 Experimental results

The presented technique will now be applied to identify two spatially distributed systems, a simply-supported beam and an asymmetric cantilever plate. Both structures are excited using bonded piezoelectric actuators. Although the simply-supported beam is easily modeled using analytic methods (albeit with experimental tuning), applying such techniques to the plate is significantly more difficult. The problem is complicated by the irregular geometry of the plate boundary.

The experimental beam and plate apparatus are shown in Figures 4.11 and 7.11 respectively.

7.6.1 *Beam identification*

7.6.1.1 *Experimental setup*

The physical parameters of the beam and properties of the piezoelectric transducer can be found in Tables 4.1 and 4.2.

Table 7.1 Identification Parameters

Frequency Range	10-200 (Hz)
Equidistant F Samples	3031
Spatial Sampling interval	2.5 cm
Identification Samples	13
Validation Samples	13
Excitation	Colored Noise

Colored noise is applied to the actuator and the spatial response is measured sequentially using a Polytec scanning laser vibrometer. Details of the data set are given in Table 7.1.

7.6.1.2 *Spatial functions*

The extracted mode shape and feed-through function samples together with their spline and linear reconstructions are shown in Figures 7.5, 7.6 and 7.7. It can be observed that the identified feed-though function is similar to that derived analytically, shown in Figure 7.3.

The Shannon reconstructed mode shapes are significantly distorted by the combination of truncation error and sample noise. It is interesting to note the effect of piezoelectric stiffness on the mode shapes of the beam shown in Figure 7.6. The length of the beam bonded to the piezoelectric patch is obviously more restricted in its deflection. Such structures with localized changes in stiffness are very difficult to model in closed form using present analytic techniques.

7.6.1.3 *Spatial response*

To evaluate model quality, we will compare the measured spatial beam response plotted in Figure 7.8 to the identified model response plotted in Figure 7.9. Each point-wise frequency response is measured from the applied actuator voltage (in volts) to the resulting displacement (in meters). A separate interlaced set of 13 points was used to perform the validation. The magnitude response of the error system, $G_y(j\omega, r) - \hat{G}_y(j\omega, r)$, where $\hat{G}_y(j\omega, r)$ denotes the model response, is plotted in Figure 7.10. In the frequency domain, the identified model is observed to accurately represent the physical system.

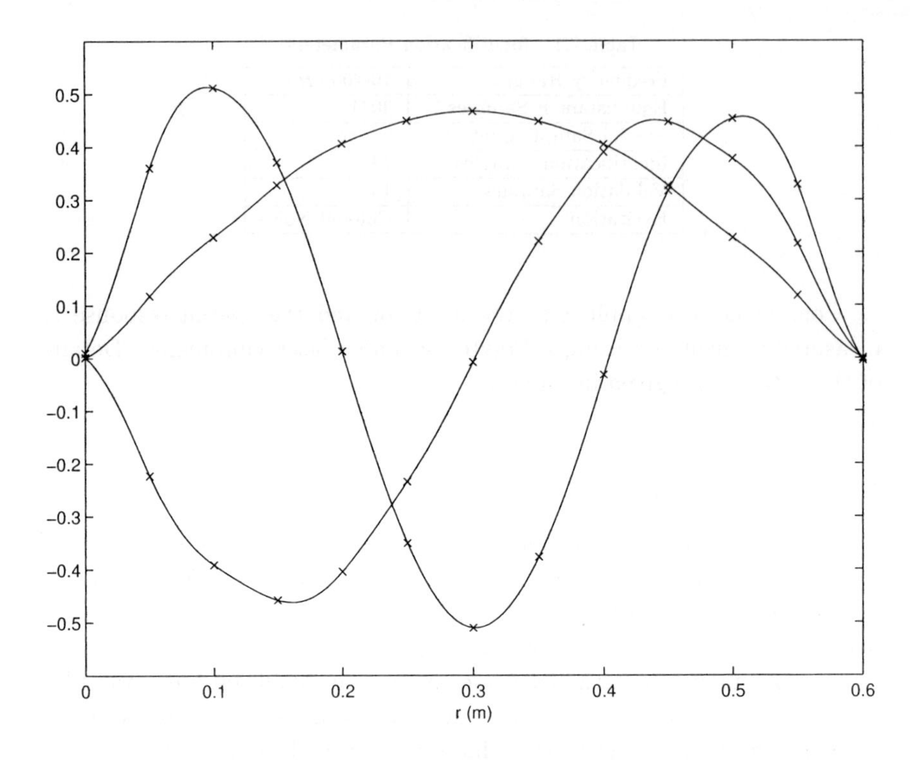

Fig. 7.5 The extracted mode samples (×) and linear reconstruction

Table 7.2 Plate Identification Parameters

Frequency Range	10-100 (Hz)
Equidistance F Samples	577
Number of Spatial Samples	468
Spatial Sampling interval	2.63 cm
Excitation	Colored Chirp

7.6.2 *Plate identification*

7.6.2.1 *Experimental setup*

The experimental plate is constructed from aluminum of 4 mm thickness. Figure 7.11 shows the experimental plate, clamped vertically by its bottom edge to an optical table. Geometry and dimensions are shown in Figure 7.12. System identification parameters are given in Table 7.2.

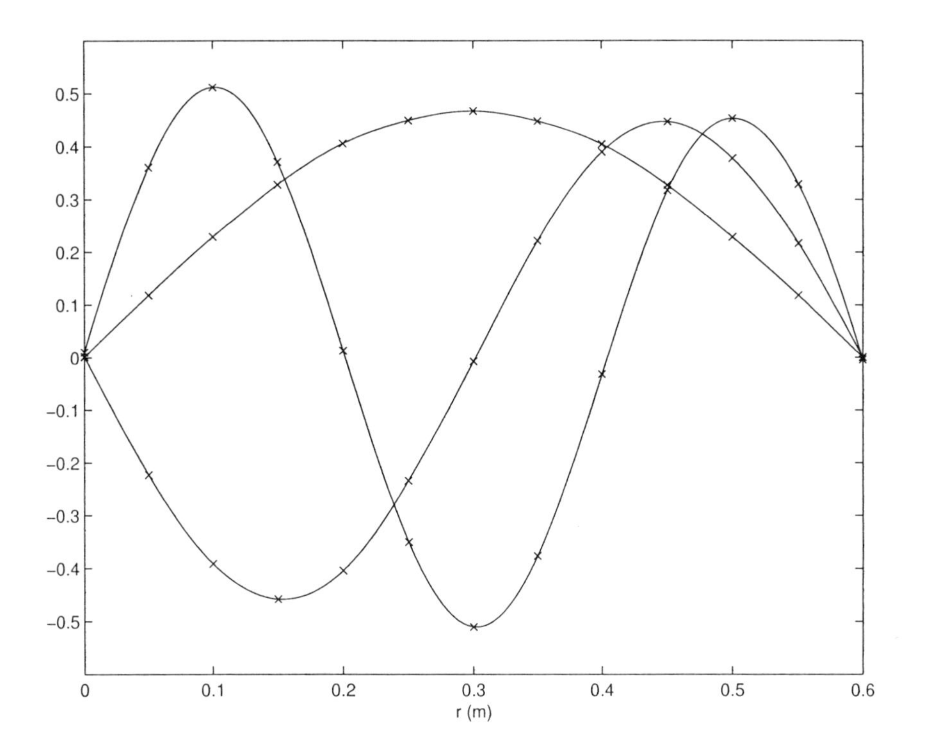

Fig. 7.6 The extracted mode samples (×) and spline reconstruction

7.6.2.2 *Spatial functions*

An estimate for the system matrix **A** is first obtained using a scattered subset of the spatial frequency samples. The location of subset points is shown in Figure 7.13. Equation (7.38) is solved to identify the mode shapes and feed-through function. The resulting mode shapes and feed-through function are plotted in Figures 7.14 and 7.15.

7.6.2.3 *Spatial response*

Due to the difficulties in visualizing a four-dimensional quantity, we evaluate model quality by taking a planar section of the spatial frequency response. An elevation of the section is shown in Figure 7.13. The measured, identified model, and error system $G_y(j\omega, r) - \hat{G}_y(j\omega, r)$ frequency responses are shown in Figures 7.16, 7.17 and 7.18 respectively. Each point-wise frequency response is measured from the applied actuator voltage to the resulting displacement (in meters). As shown by the magnitude of the er-

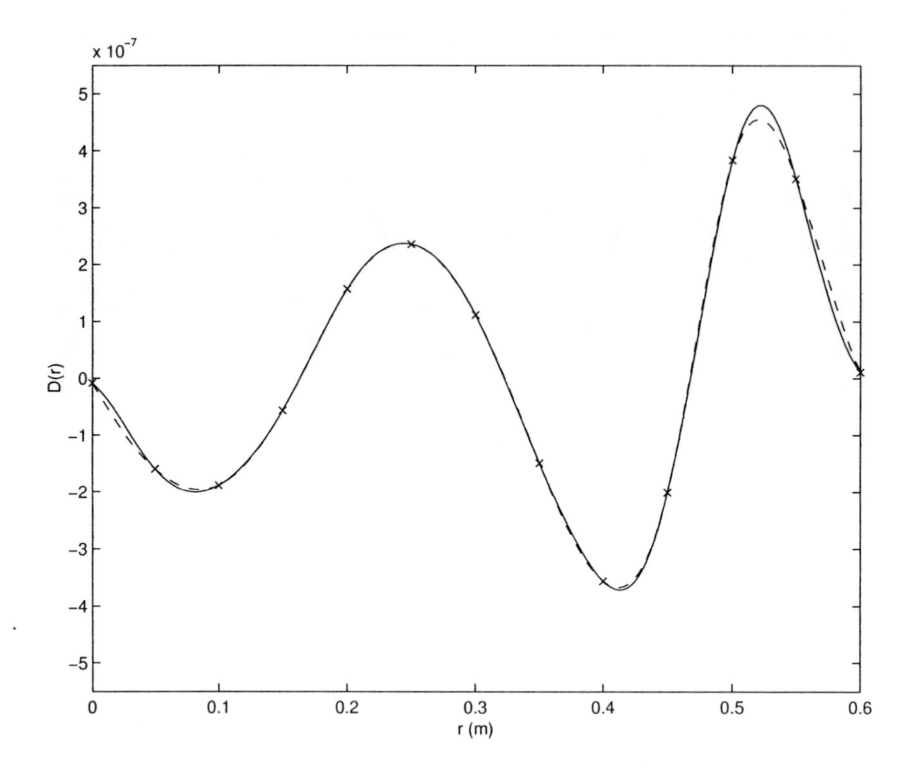

Fig. 7.7 The extracted feed-through function samples ($\times$), the linear reconstruction (–), and spline reconstruction (- -)

ror system response, in the frequency domain, a good correlation between the experimental data and model response can be observed.

7.7 Conclusions

A technique has been presented for identifying a class of distributed parameter systems from a set of spatially distributed frequency responses. The systems are modeled as a finite sum of second order transfer functions with spatially variant numerators and a feed-through term.

In an attempt to evenly distribute model error, the identification is cast as a single-input multi-output identification problem. An estimate for the system dynamics is sought using a frequency domain subspace algorithm. Samples of the mode shapes and feed-through function are then identified and used to reconstruct the continuous functions. If the spatial Fourier

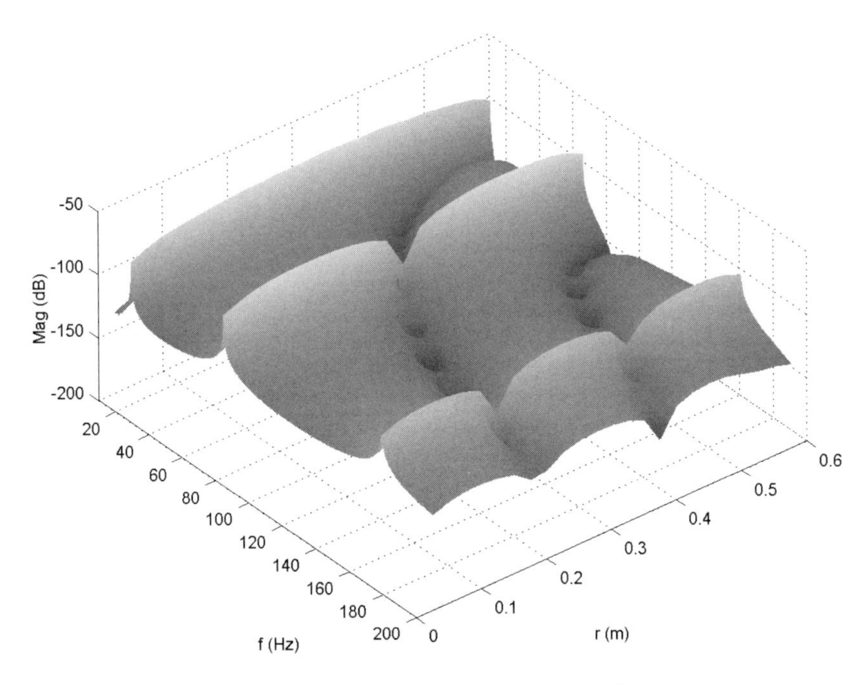

Fig. 7.8 The experimental beam spatial frequency response from an applied actuator voltage to the measured displacement $G_y(j\omega, r)$

transform is known, the error due to under sampling can be quantified.

Experimental identification of a simply-supported beam and cantilever plate has shown an adequate correlation in the frequency domain between the measured system and identified model. In both cases the majority of discrepancy is due to small errors in the resonance frequencies. It is anticipated that future contributions in this area will involve the development of an efficient optimization algorithm to minimize such errors.

Other outstanding problems to date include: the automatic identification of non-distributed dynamics $H(s)$, experimental identification incorporating piezoelectric sensor voltages, time domain identification techniques and stochastic analysis.

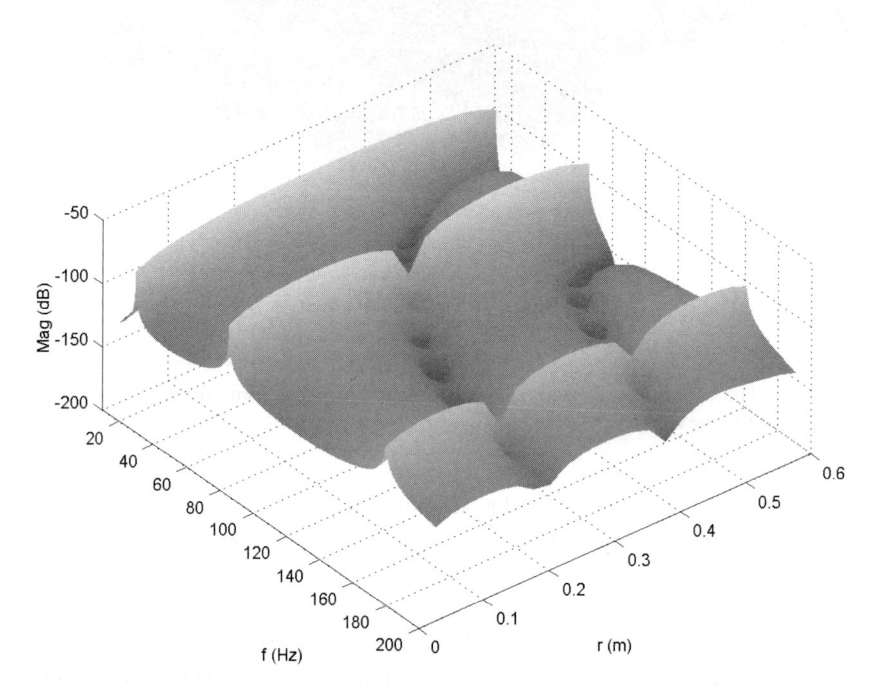

Fig. 7.9 The spline reconstructed model response, from an applied actuator voltage to the measured displacement $\hat{G}_y(j\omega, r)$

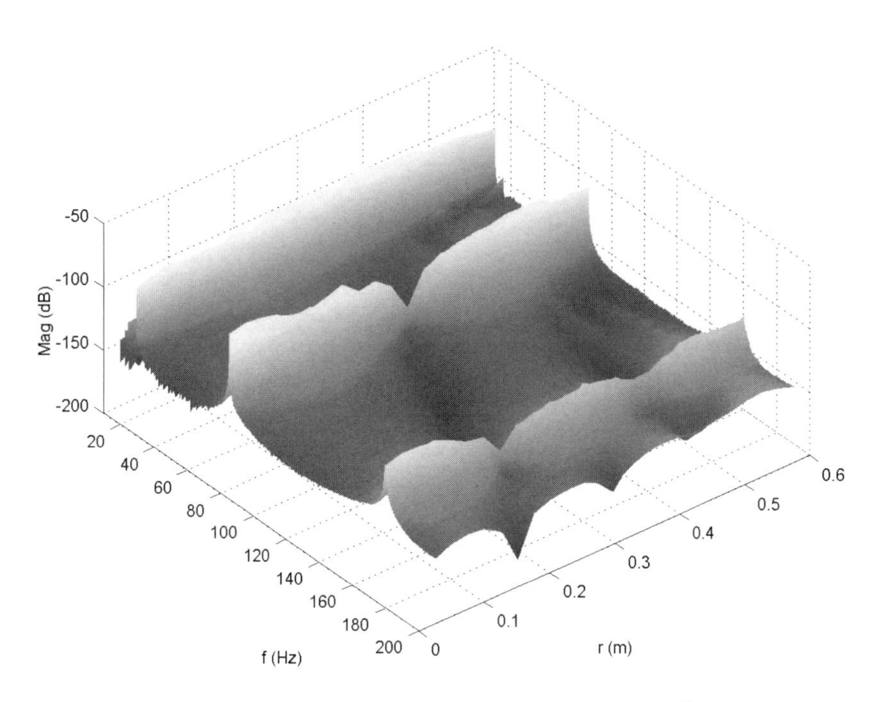

Fig. 7.10 The error system response response, $G_y(j\omega, r) - \hat{G}_y(j\omega, r)$

Fig. 7.11 Experimental plate apparatus

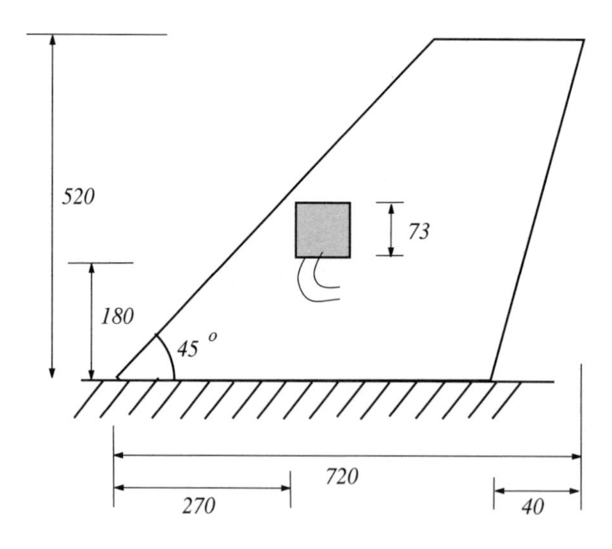

Fig. 7.12 Plate geometry (mm)

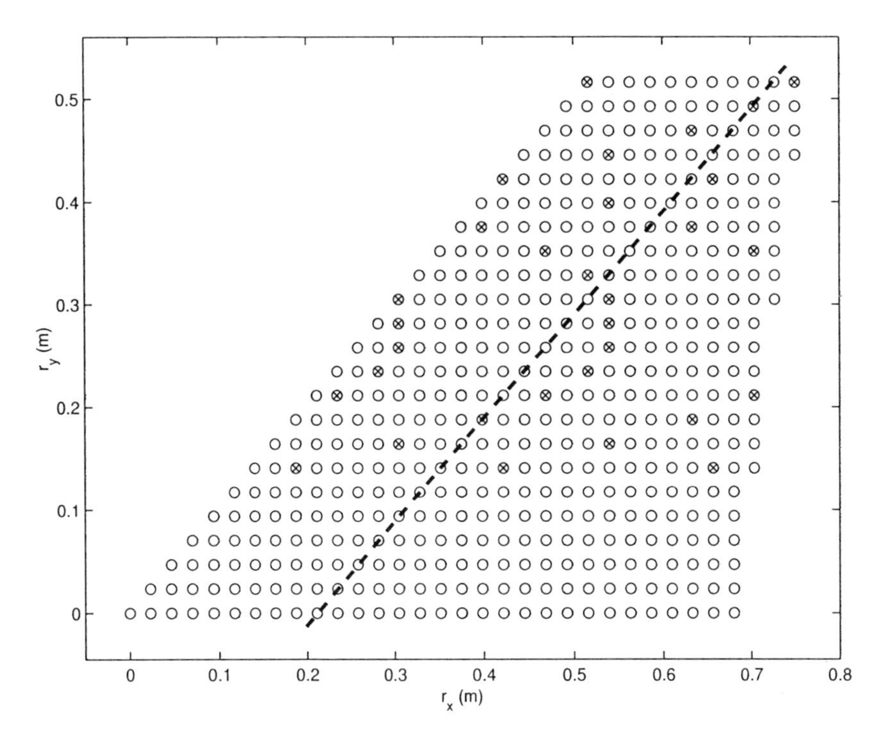

Fig. 7.13 Distribution of the spatial samples. An '×' represents the location of a sample used to identify the system matrix **A**. The dashed line represents the side elevation of a spatial frequency response cross section used to analyze model quality

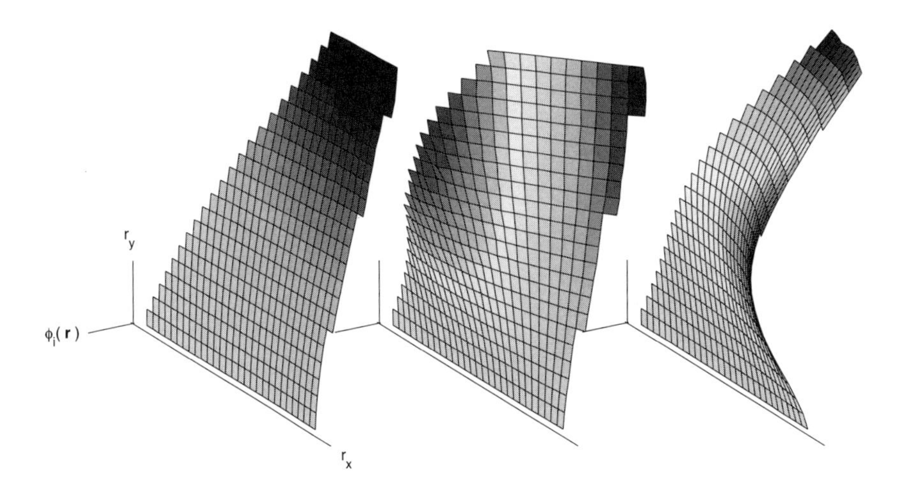

Fig. 7.14 The normalized first, second, and third modes of the cantilever fin

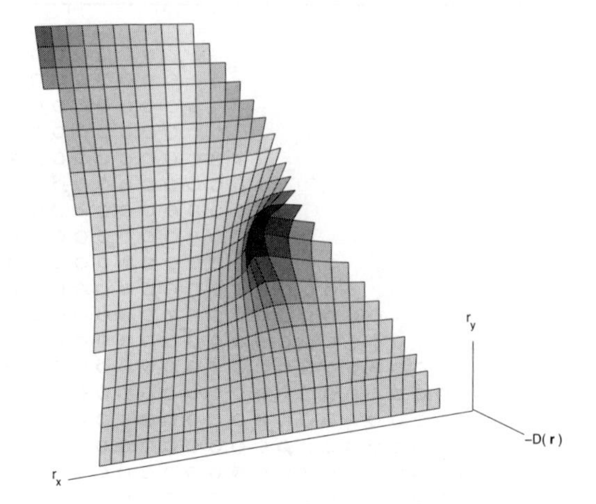

Fig. 7.15 The identified feed-through function $D(\mathbf{r})$

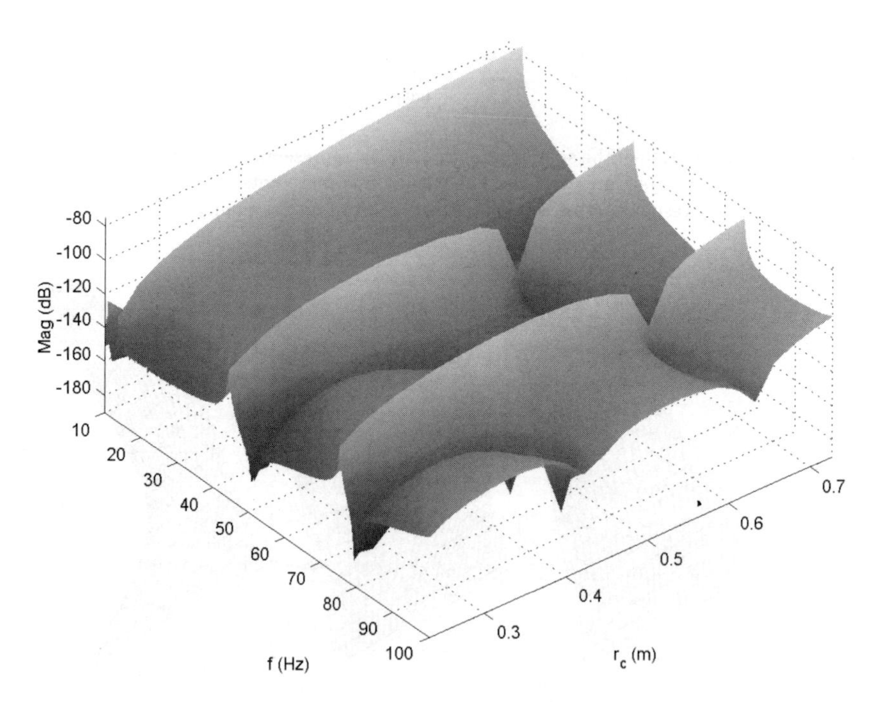

Fig. 7.16 A cross section of the cantilever plate measured spatial frequency response. The response is measured from the applied actuator voltage to the resulting displacement

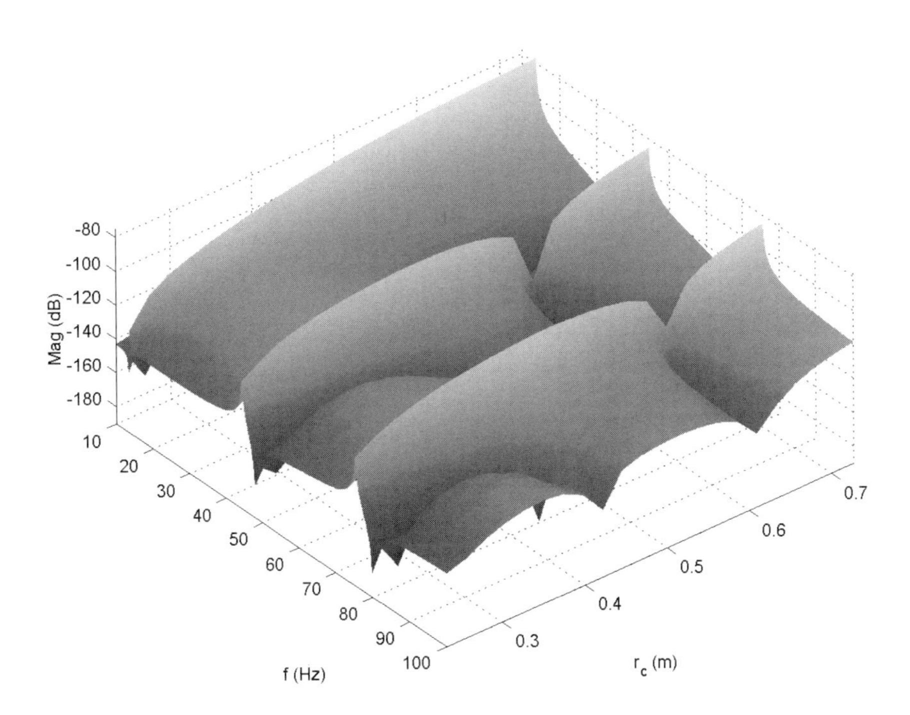

Fig. 7.17 A cross section of the spline reconstructed model response

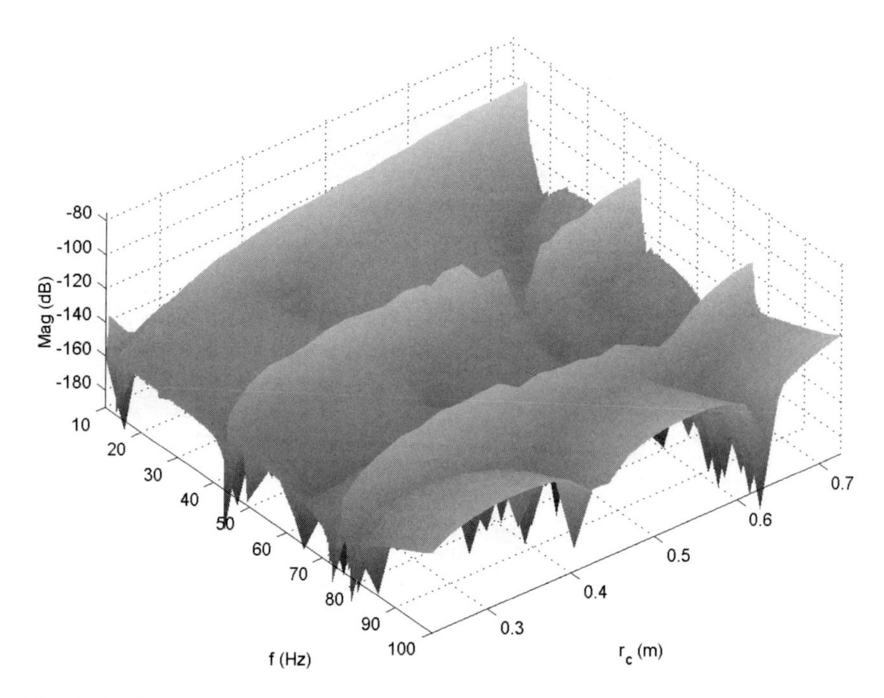

Fig. 7.18 A cross section of the plate error system frequency response, $G_y(j\omega, r) - \hat{G}_y(j\omega, r)$

Appendix A

Frequency domain subspace system identification

A.1 Introduction

The identification of a transfer function from frequency response data has been widely addressed in the literature [61, 88, 96, 89]. In the traditional way, a system is modeled as the ratio of two real polynomials, the coefficients are determined by minimizing some non-linear least squares criterion. The solution is sought by either iterative numerical search or solution of a series of least squares SK iterations. Neither method is guaranteed to converge to the global minima. A further disadvantage is the sensitivity of pole-zero locations to polynomial factorization of high order systems. This can be reduced by introducing other parameterizations such as orthonormal polynomials [82].

Alternatively, frequency domain state space methods have become available. The first to appear in this category were algorithms based on the famous Ho and Kalman realization [40] or Kung's smoothed version [51]. The impulse response of the system, which can be estimated from the IDFT of the data, is used to find a minimal state-space realization [49]. The frequency domain data must be evenly spaced to allow computation of the IDFT.

A frequency domain counterpart to the time domain subspace method developed by De Moor and Vandewalle [79] was presented by Liu and coworkers [60]. This method does not require equidistant frequency samples. An extension by McKelvy et al. [65] is shown to provide strongly consistent estimates if the noise model is known (coloring or covariance). This work utilizes portions of the frequency domain subspace identification algorithm [60], and later [65].

A.2 Frequency Domain Subspace Algorithm

A.2.1 *Preliminaries*

Consider the stable time-invariant discrete linear system of order $2N$ in state space form

$$\begin{aligned} \mathbf{x}(k+1) &= \mathbf{A}\mathbf{x}(k) + \mathbf{B}\mathbf{u}(k) \\ \mathbf{y}(k) &= \mathbf{C}\mathbf{x}(k) + \mathbf{D}\mathbf{u}(k), \end{aligned} \tag{A.1}$$

where $\mathbf{u}(k) \in \mathbf{R}^Q$, $\mathbf{y}(k) \in \mathbf{R}^L$ and $\mathbf{x}(k) \in \mathbf{R}^{2N}$. By considering real valued signals we implicitly assume that the quadruple $[\mathbf{A}, \mathbf{B}, \mathbf{C}, \mathbf{D}]$ is also real. We also assume the realization is minimal which implies the system is both observable and controllable. Note that all systems having the same input/output relationship are related by a non-singular similarity transform $\mathbf{T}$, where $\left[\mathbf{T}^{-1}\mathbf{A}\mathbf{T}, \mathbf{C}\mathbf{T}, \mathbf{T}^{-1}\mathbf{B}, \mathbf{D}\right]$ is the equivalent state space realization.

The discrete time Fourier transform of a system is defined as

$$\mathcal{F}\left\{f(k)\right\} = F(j\omega) = \sum_{k=-\infty}^{\infty} f(k)e^{-j\omega k}. \tag{A.2}$$

Applying the transform to (A.1) yields

$$\begin{aligned} e^{j\omega}\mathbf{X}(j\omega) &= \mathbf{A}\mathbf{X}(j\omega) + \mathbf{B}\mathbf{U}(j\omega) \\ \mathbf{Y}(j\omega) &= \mathbf{C}\mathbf{X}(j\omega) + \mathbf{D}\mathbf{U}(j\omega). \end{aligned} \tag{A.3}$$

By eliminating the state, we obtain

$$\mathbf{Y}(j\omega) = \mathbf{G}(e^{j\omega})\mathbf{U}(j\omega), \tag{A.4}$$

where $\mathbf{G}(e^{j\omega}) = \mathbf{D} + \mathbf{C}(s\mathbf{I} - \mathbf{A})^{-1}\mathbf{B} + \mathbf{D}$ is the system transfer function.

A.2.2 *Identification of the System Matrix*

An algorithm will be presented that is applicable for non-uniformly spaced frequency data

$$G_k = G(e^{j\omega_k}) + n_k \; k = 1, \cdots, M. \tag{A.5}$$

We now define some matrices used in the presentation of the algorithm.

$X^i(\omega)$ the resulting state transform when $\mathbf{U}(j\omega) = e_i$ the unit vector with one in the ith position. Define the compound state matrix

$$X^c(j\omega) = \left[X^1 \; X^2 \cdots \; X^Q \right] \in \mathbf{C}^{2N \times Q} \tag{A.6}$$

The ordered collection of all component states

$$\mathbf{X}^c = \frac{1}{\sqrt{M}} \left[X^c(j\omega_1) \; X^c(j\omega_2) \cdots \; X^c(j\omega_M) \right] \in \mathbf{C}^{2N \times QM} \tag{A.7}$$

The extended observability matrix with q block rows

$$\mathcal{O}_q = \begin{bmatrix} \mathbf{C} \\ \mathbf{CA} \\ \vdots \\ \mathbf{CA}^{(q-1)} \end{bmatrix} \in \mathbf{R}^{qL \times 2N} \tag{A.8}$$

The lower triangular Toeplitz matrix

$$\Gamma = \begin{bmatrix} \mathbf{D} & \mathbf{0} & \cdots & 0 \\ \mathbf{CB} & \mathbf{D} & \cdots & 0 \\ \vdots & \vdots & \ddots & \vdots \\ \mathbf{CA}^{(q-2)}\mathbf{B} & \mathbf{CA}^{(q-3)}\mathbf{B} & \cdots & \mathbf{D} \end{bmatrix} \in \mathbf{R}^{qL \times qQ} \tag{A.9}$$

The frequency response matrix

$$\mathbf{G} = \frac{1}{\sqrt{M}} \begin{bmatrix} G_1 & G_2 & \cdots & G_M \\ e^{j\omega_1} G_1 & e^{j\omega_2} G_2 & \cdots & e^{j\omega_M} G_M \\ e^{j2\omega_1} G_1 & e^{j2\omega_2} G_2 & \cdots & e^{j2\omega_M} G_M \\ \vdots & \vdots & \ddots & \vdots \\ e^{j(q-1)\omega_1} G_1 & e^{j(q-1)\omega_2} G_2 & \cdots & e^{j(q-1)\omega_M} G_M \end{bmatrix} \in \mathbf{C}^{qL \times QM} \tag{A.10}$$

The noise matrix

$$\mathbf{V} = \frac{1}{\sqrt{M}} \begin{bmatrix} v_1 & v_2 & \cdots & v_M \\ e^{j\omega_1} v_1 & e^{j\omega_2} v_2 & \cdots & e^{j\omega_M} v_M \\ e^{j2\omega_1} v_1 & e^{j2\omega_2} v_2 & \cdots & e^{j2\omega_M} v_M \\ \vdots & \vdots & \ddots & \vdots \\ e^{j(q-1)\omega_1} v_1 & e^{j(q-1)\omega_2} v_2 & \cdots & e^{j(q-1)\omega_M} v_M \end{bmatrix} \in \mathbf{C}^{qL \times QM} \tag{A.11}$$

The block Vandermonde matrix

$$\mathbf{W}_Q = \frac{1}{\sqrt{M}} \begin{bmatrix} \mathbf{I}_Q & \mathbf{I}_Q & \cdots & \mathbf{I}_Q \\ e^{j\omega_1}\mathbf{I}_Q & e^{j\omega_2}\mathbf{I}_Q & \cdots & e^{j\omega_M}\mathbf{I}_Q \\ e^{j2\omega_1}\mathbf{I}_Q & e^{j2\omega_2}\mathbf{I}_Q & \cdots & e^{j2\omega_M}\mathbf{I}_Q \\ \vdots & \vdots & \ddots & \vdots \\ e^{j(q-1)\omega_1}\mathbf{I}_Q & e^{j(q-1)\omega_2}\mathbf{I}_Q & \cdots & e^{j(q-1)\omega_M}\mathbf{I}_Q \end{bmatrix} \in \mathbf{C}^{qL \times QM} \tag{A.12}$$

We can describe the transfer function in state-space form

$$\begin{aligned} e^{j\omega}\mathbf{X}^c(j\omega) &= \mathbf{A}\mathbf{X}^c(j\omega) + \mathbf{B} \\ \mathbf{G}(e^{j\omega}) &= \mathbf{C}\mathbf{X}^c(j\omega) + \mathbf{D} \end{aligned} \tag{A.13}$$

By recursive use of (A.13) and (A.8) we obtain the relation

$$\begin{bmatrix} G(e^{j\omega}) \\ e^{j\omega}G(e^{j\omega}) \\ \vdots \\ e^{j(q-1)\omega}G(e^{j\omega}) \end{bmatrix} = \mathcal{O}\mathbf{X}^c(j\omega) + \Gamma \begin{bmatrix} \mathbf{I}_Q \\ e^{j\omega}\mathbf{I}_Q \\ \vdots \\ e^{j(q-1)\omega}\mathbf{I}_Q \end{bmatrix}. \tag{A.14}$$

in matrix notation

$$\mathbf{G} = \mathcal{O}\mathbf{X}^c + \Gamma\mathbf{W}_Q + \mathbf{V}. \tag{A.15}$$

Since we assume the system is minimal, $(\mathbf{A}, \mathbf{B})$ is a controllable pair. Lemma 2 of [65] shows that $\mathbf{X}^c$ is of full rank $2N$, therefore the matrix product $\mathcal{O}\mathbf{X}^c$ has range space equal to $\mathcal{O}$. Since $\mathcal{O}$ is real, we are interested in the real range space of $\mathcal{O}\mathbf{X}^c$.

Equation (A.15) can be rewritten as a real valued relation

$$\overbrace{\left[Re\,\mathbf{G}\ Im\,\mathbf{G}\right]}^{\mathcal{G}} = \mathcal{O}\overbrace{\left[Re\,\mathbf{X}^{c\}}\ Im\,\mathbf{X}^c\right]}^{\mathcal{X}} + \Gamma\overbrace{\left[Re\,\mathbf{W}_Q\ Im\,\mathbf{W}_Q\right]}^{\mathcal{W}} + \overbrace{\left[Re\,\mathbf{V}\ Im\,\mathbf{V}\right]}^{\mathcal{V}}, \tag{A.16}$$

where $\mathbf{W}_Q$, $\mathbf{X}^c$, $\mathcal{W}$, $\mathcal{X}$ are of full rank [65].

Consider the noise free case, if $\Gamma\mathcal{W}$ were also equal to zero, the range space of $\mathcal{G}$ would equal the range space of $\mathcal{O}$ if $q > 2N$ and $M \geq 2N$. A factorization would yield an observability matrix similar to $\mathcal{O}$.

The key subspace technique is to multiply (A.15) from right to eliminate the term $\Gamma\mathcal{W}$. One such matrix is the projection onto the null space of $\mathcal{W}$, given by

$$\mathcal{W}^{\perp} = \mathbf{I} - \mathcal{W}^T(\mathcal{W}\mathcal{W}^T)^{-1}\mathcal{W}. \tag{A.17}$$

Equation (A.15) then simplifies to

$$\mathcal{G}\mathcal{W}^{\perp} = \mathcal{O}\mathcal{X}\mathcal{W}^{\perp}, \tag{A.18}$$

where the range space of $\mathcal{G}\mathcal{W}^{\perp}$ equals the range space of $\mathcal{O}$ unless rank cancellation occurs. If the frequencies are distinct, there are sufficient data samples $M \geq q + 2N$ and $(\mathbf{A}, \mathbf{B})$ is a controllable pair, the row spaces of $\mathbf{X}^c$ and $\mathbf{W}_Q$ do not intersect and the range space of $\mathcal{G}\mathcal{W}^{\perp}$ equals the range space of $\mathcal{O}$ [65].

The extended observability matrix $\mathcal{O}$, which precisely spans the column space of $\mathcal{O}\mathcal{X}\mathcal{W}^{\perp}$, can now be recovered by a singular value decomposition, i.e.

$$\mathcal{O}\mathcal{X}\mathcal{W}^{\perp} = \begin{bmatrix} U_s \, U_o \end{bmatrix} \begin{bmatrix} \Sigma_s & 0 \\ 0 & \Sigma_o \end{bmatrix} \begin{bmatrix} V_s \\ V_o \end{bmatrix}, \tag{A.19}$$

where $U_s \in \mathbf{R}^{2L \times 2N}$ contains the $2N$ principle singular vectors and Σ_s contains the corresponding singular values. In the noise free case, Σ_o will be zero and there exists a non-singular matrix $\mathbf{T} \in \mathbf{R}^{2N \times 2N}$ such that

$$\mathcal{O} = U_s \mathbf{T}. \tag{A.20}$$

U_s is the observability matrix for some realization of the identified system. The system matrix is easily extracted from $\mathcal{O}$

$$\mathcal{O}(\, L+1 : qL \,,\, 1 : 2N \,) = \mathcal{O}(\, 1 : (q-1)L \,,\, 1 : 2N \,)\widehat{\mathbf{A}}. \tag{A.21}$$

A.2.3 *Continuous Time Conversion*

The problem of parameter estimation is better conditioned in the discrete domain as powers of $e^{j\omega}$ form a natural orthonormal basis [65]. If the standard zero order hold approximations are valid, the continuous time system matrix may be found indirectly via the bilinear transform.

$$\mathbf{A}_g = \frac{2}{T}(\mathbf{I} + \mathbf{A})^{-1}(\mathbf{A} - \mathbf{I}), \tag{A.22}$$

where the parameter T is free to be varied under the condition that $\frac{2}{T}$ is not a pole of the continuous time system.

A.2.4 *Summary*

By replacing the general frequency data with the spatial response, the system matrix and hence dynamics of each mode can be found. Each sub-column of the frequency response matrix now contains the response from one of the system inputs to the displacement at each measured point on the spatial domain.

Bibliography

[1] T. E. Alberts, T. V. DuBois, and H. R. Pota. Experimental verification of transfer functions for a slewing piezoelectric laminate beam. *Control Engineering Practice*, 3(2):163–170, 1995.

[2] T. E. Alberts and H. R. Pota. Broadband dynamic modification using feedforward control. In *Proc.eedings of the 1995 Design Engineering Technical Conference*, volume 3, Part B, pages 735–744, Boston, 17–21 September 1995. ASME.

[3] H.T. Banks, R.C. Smith, and Y. Wang. *Smart Material Structures: Modeling, Estimation and Control.* Wiley - Masson, Chichester - Paris, 1996.

[4] F.P. Beer and E.R. Johnston, Jr. *Mechanics of Materials.* McGraw-Hill, London, 1992. 2nd ed., collaboration with J.T. DeWolf.

[5] R. L. Bisplinghoff and H. Ashley. *Principles of Aeroelasticity.* Dover Publications Inc., 1962.

[6] T. Blu and M. Unser. Quantitative fourier analysis of approximation techniques: Part 1 - interpolators and projectors. *IEEE Transactions on Signal Processing*, 47(10):2783–2795, October 1999.

[7] S.F. Borg. *Fundamentals of Engineering Elasticity.* Van Nostrand, Princeton, 1962.

[8] J. R. Carstens. *Electrical Sensors and Transducers.* Prentice-Hall, 1993.

[9] T.R. Chandrupatla and A.D. Belegundu. *Introduction to Finite Elements in Engineering.* Prentice Hall, Upper Saddle River, N.J., 1997. 2nd ed.

[10] Y.K. Cheung and A.Y.T. Leung. *Finite Element Methods in Dynamics.* Science Press; Kluwer Academic Publishers, Beijing, New York; Dordrecht, Boston, 1991.

[11] R. L. Clark. Accounting for out-of-bandwidth modes in the assumed modes approach: implications on colocated output feedback control. *Transactions of the ASME, Journal of Dynamic Systems, Measurement, and Control*, 119:390–395, 1997.

[12] R.L Clark, W.R. Saunders, and G.P. Gibbs. *Adaptive Structures: Dynamics and Control.* Wiley, Canada, 1998.

[13] R. D. Cook. *Finite Element Modelling for Stress Analysis.* John Wiley and Sons, 1995.

[14] E.F. Crawley and J de Luis. Use of piezoelectric actuators as elements of intelligent structures. *AIAA Journal*, pages 1373–1385, October 1987.

[15] H. H. Cudney. *Distributed structural control using multilayered piezoelectric actuators.* PhD thesis, SUNY Buffalo, NY, 1989.

[16] C.W. de Silva. *Vibration: Fundamentals and Practice.* CRC Press, Boca Raton, 2000.

[17] M.A. Demetriou. A numerical algorithm for the optimal placement of actuators and sensors for flexible structures. In *Proceedings of the American Control Conference*, pages 2290–2294, Chicago, Illinois, USA, June 2000.

[18] E.K. Dimitriadis, C.R. Fuller, and C.A Rogers. Piezoelectric actuators for distributed vibration excitation of thin plates. *ASME Journal of Vibration and Acoustics*, 113:100–107, January 1991.

[19] D. J. Ewins. Modal testing as an aid to vibration analysis. In *Proc. Conference on Mechanical Engineering*, May 1990.

[20] F. Fahroo and Y. Wang. Optimal location of piezoceramic actuators for vibration suppression of a flexible structure. In *Proceedings of the* 36th *IEEE Conference on Decision & Control*, pages 1966–1971, San Diego, CA, USA, December 1997.

[21] K. D. Fampton and R. L. Clark. Active control of panel flutter with linearized potential flow. In *Proceedings of the AIAA/ASME/ASCE/AHS/ASC 36th Structures, Structural Dynamics, and Matrials Conference*, pages 2273–2280, New Orleans, LA, 1995.

[22] R. L. Fante. *Signal Analysis and Estimation. An Introduction.* John Wiley and Sons, 1988.

[23] A. R. Fraser and R. W. Daniel. *Pertubation Techniques for Flexible Manipulators.* Kluwer Academic Publishers, 1991.

[24] M.I. Friswell. *Finite Element Model Updating in Structural Dynamics.* Kluwer Academic Publishers, Dordrecht, Boston, 1995.

[25] C.R. Fuller, S.J. Elliot, and P.A. Nelson. *Active Control of Vibration.* Academic Press, London, 1996.

[26] B. D. O. Anderson G. Obinata. *Model Reduction for Control System Design.* Springer-Verlag, 2000.

[27] W.K. Gawronski. Actuator and sensor placement for structural testing and control. *Journal of Sound and Vibration*, 208(1):101–109, 1997.

[28] W.K. Gawronski. *Dynamics and Control of Structures: A Modal Approach.* Springer, New York, 1998.

[29] J.M. Gere and S.P. Timoshenko. *Mechanics of Materials.* Chapman & Hall, London, 1991. 3rd ed.

[30] K. Glover. All optimal Hankel-norm approximations of linear multivariable systems and their l^{∞}-error bounds. *International Journal of Control*, 39(6):1115–1193, 1984.

[31] I. S. Gradsteyn and I. M. Ryshik. *Table of Integrals, Series, and Products.* Academic Press, 4th edition, 1994.

[32] M. Green and D. J. N. Limebeer. *Linear Robust Control.* Prentice Hall, 1994.

[33] D. Halim and S. O. R. Moheimani. An optimization approach to optimal placement of collocated piezoelectric actuators and sensors on a thin plate. Accepted for publication in *Mechatronics*.

[34] D. Halim and S. O. R. Moheimani. Experiments in spatial H_{∞} control of a piezoelectric laminate beam. In S. O. R. Moheimani, editor, *Perspectives in Robust Control.* Springer-Verlag, London, 2001.

[35] D. Halim and S. O. R. Moheimani. Spatial control of flexible structures - application of spatial h-infinity control to a piezoelectric laminate beam. In *Proc. American Control Conference*, Arlington, Virginia, June 2001.

[36] D. Halim and S. O. R. Moheimani. Spatial resonant control of flexible structures - application to a piezoelectric laminate beam. *IEEE Transactions on Control System Technology*, 9(1):37–53, January 2001.

[37] D. Halim and S. O. R. Moheimani. Experimental implementation of spatial H_∞ control on a piezoelectric laminate beam. September 2002.

[38] D. Halim and S. O. R. Moheimani. Spatial H_2 control of a piezoelectric laminate beam: Experimental implementation. *IEEE Transactions on Control Systems Technology*, 10(4):533–546, July 2002.

[39] J. Heng, J. C. Akers, R. Venugopal, M. Lee, A. G. Sparks, P. D. Washabaugh, and D. Bernstien. Modeling, identification, and feedback control of noise in an acoustic duct. *IEEE Transactions on Control Systems Technology*, 4(3):283–291, 1996.

[40] B. L. Ho and R. E. Kalman. Effective construction of linear state variable models from Input/Output functions. *Regelungstechnik*, 14(12):545–548, 1966.

[41] J. Hong, J. C. Akers, R. Venugopal, M. Lee, A. G. Sparks, P. D. Washabaugh, and D. Bernstein. Modeling, identification, and feedback control of noise in an acoustic duct. *IEEE Transactions on Control Systems Technology*, 4(3):283–291, May 1996.

[42] R. Hummel. Sampling for spline reconstruction. *SIAM Journal of Applied Mathematics*, 43(2):278–288, April 1983.

[43] J.K. Hwang, C.H. Choi, C.K. Song, and J.M. Lee. Robust LQG control of an all-clamped thin plate with piezoelectric actuators/ sensors. *IEEE/ ASME Transactions on Mechatronics*, 2(3):205–212, September 1997.

[44] K.U. Ingard. *Fundamental of Waves and Oscillations*. Cambridge, Cambridge, 1988.

[45] D. J. Inman. *Engineering Vibration*. Prentice Hall, 2nd edition, 2001.

[46] M.H. Jawad. *Theory and Design of Plate and Shell Structures*. Chapman & Hall, New York, 1994.

[47] A. J. Jerri. The shannon sampling theorem - its various extensions and applications: A tutorial review. *Proceedings of the IEEE*, 65(11):1565–1596, November 1977.

[48] S.M Joshi. *Control of Large Flexible Space Structures*. Springer, Berlin, 1989.

[49] J. N. Juang and H. Suzuki. An eigensystem realization algorithm in frequency domain for modal parameter identification. *Journal of Vibration, Acoustics, Stress, and Reliability*, 110:24–29, January 1988.

[50] A. Krzyzak and E Rafajłowicz M. Pawlak. Moving average restoration of bandlimited signals from noisy observations. *IEEE transactions on signal processing*, 45(12):2967–2976, December 1997.

[51] S. Y. Kung. A new identification and model reduction algorithm via singular value decomposition. In *Proc. Conference on Circuits, Systems, and Computation*, pages 705–714, Pacific Grove, CA, 1978.

[52] H. Kwakernaak and R. Sivan. *Modern Signals and Systems*. Prentice Hall, Englewood Cliffs, N.J., 1991.

[53] P. Lancaster and K. Salkauskas. *Curve and Surface Fitting*. Academic Press, 1986.

[54] K. B. Lazarus and E. F. Crawley. Multivariable high-authority control of plate-like active structures. In *Proceedings of the AIAA/ASME/ASCE/AHS/ASC 33rd Structures, Structural Dynamics, and Matrials Conference*, pages 931–945, Dallas, TX, 1992.

[55] K. B. Lazarus and E. F. Crawley. Mutlivariable active lifting surface control using strain actuation: analytical and experimental results. In *Proc. of the 3rd International Conference on Adaptive Structures*, pages 87–101, San Diego, CA, 1992.

[56] C. K. Lee. *Piezoelectric Laminates for Torsional and Bending Modal Control: Theory and Experiment.* PhD thesis, Cornell University, 1987.

[57] C. K. Lee and F. C. Moon. Modal sensors/actuators. *ASME Journal of Applied Mechanics*, 57:434–441, June 1990.

[58] F. L. Lewis. *Applied Optimal Control and Estimation.* Prentice Hall, 1992.

[59] Y. Lim, V. V. Varadan, and V. K. Varadan. Closed-loop finite element modelling of active/passive damping in structural vibration control. In *Proc. SPIE Smart Materials and Structures, Mathematics and Control in Smart Structures, SPIE Vol.3039*, San Diego, CA, March 1997.

[60] K. Liu, R. N. Jacques, and D. W. Miller. Frequency domain structural system identification by observability range space extraction. In *Proc. American Control Conference, Vol.1*, pages 107–111, Baltimore, MD, June 1994.

[61] L. Ljung. *System Identification: Theory for the User.* Prentice Hall, 1999.

[62] N. M. Medes Maia and J. M. Montalvao e Silva, editors. *Theoretical and Experimental Modal Analysis.* Research Studies Press, Hertfordshire, England, 1997.

[63] E.H. Mansfield. *The Bending and Stretching of Plates.* Pergamon Press, Oxford, 1964.

[64] T. McKelvey, A. J. Fleming, and S. O. Reza Moheimani. Subspace based system identification for an acoustic enclosure. *ASME Jounrnal of Vibration and Acoustics*, 2002.

[65] T. McKelvy, H. Akcay, and L. Ljung. Subspace based multivariable system identification from frequency response data. *IEEE Transactions on Automatic Control*, 41(7):960–978, July 1996.

[66] T. McKelvy and L. Ljung. Frequency domain maximum likelihood identification. In *Proc. IFAC Symposium on System Identification*, pages 1741–1746, Fukuoda, Japan, July 1997.

[67] L. Meirovitch. *Elements of Vibration Analysis.* McGraw-Hill, Sydney, 2nd ed., 1986.

[68] L. Meirovitch. *Dynamics and control of structures.* John Wiley & Sons, 1990.

[69] S. O. R. Moheimani. Broadband disturbance attenuation over an entire beam. *Journal of Sound and Vibration*, 227(4):807–832, 1999.

[70] S. O. R. Moheimani. Experimental verification of the corrected transfer function of a piezoelectric laminate beam. *IEEE Transactions on Control Systems Technology*, 8(4):660–666, July 2000.

[71] S. O. R. Moheimani. Minimizing the effect of out of bandwidth modes in truncated structure models. *Transactions of the ASME - Journal of Dynamic Systems, Measurement, and Control*, 122(1):237–239, March 2000.

[72] S. O. R. Moheimani. Model correction for sampled-data models of structures. *AIAA Journal of Guidance, Control and Dynamics*, 24(3):634–637, 2001.

[73] S. O. R. Moheimani, H. R. Pota, and I. R. Petersen. Spatial balanced model reduction for flexible structures. *Automatica*, 32(2):269–277, February 1999.

[74] S. O. R. Moheimani and T. Ryall. Considerations in placement of piezoceramic actuators that are used in structural vibration control. In *Proc. IEEE Conference on Decision and Control*, pages 1118–1123, Phoenix, Arizona, USA, December 1999.

[75] S.O.R. Moheimani. Minimizing the effect of out-of-bandwidth dynamics in the models of reverberant systems that arise in modal analysis: Implications on spatial $\mathcal{H}_\infty$ control. *Automatica*, 36:1023–1031, 2000.

[76] S.O.R. Moheimani and M. Fu. Spatial $\mathcal{H}_2$ norm of flexible structures and its application in model order selection. In *Proceedings of the* 37th *IEEE Conference on Decision & Control*, pages 3623–3624, Tampa, Florida, USA, December 1998.

[77] S.O.R. Moheimani and W.P. Heath. Model correction for a class of spatio-temporal systems. *Automatica*, 38(1):147–155, 2002.

[78] S.O.R. Moheimani and T. Ryall. Considerations in placement of piezoceramic actuators that are used in structural vibration control. In *Proceedings of the* 38th *IEEE Conference on Decision & Control*, pages 1118–1123, Phoenix, Arizona, USA, December 1999.

[79] B. De Moor and J. Vanderwalle. A geometrical strategy for the identification of state space models of linear multivariate systems with singular value decomposition. In *Proc. International Symposium on Multivariable System Technology*, pages 59–69, Plymouth, UK, April 1987.

[80] A. J. Moulson and J. M. Herbert. *Electroceramics: Materials, Properties, Applications*. Chapman and Hall, London, 1990.

[81] P. A. Nelson and S. J. Elliot. *Active Control of Sound*. Academic Press, 1992.

[82] B. Ninness, H. Hjalmarsson, and F. Gustafsson. The fundamental role of orthonormal bases in system identification. *IEEE Transactions on Automatic Control*, 44(7):1384–1407, July 1999.

[83] Institute of Electrical and Electronics Engineers Inc. IEEE standard on piezoelectricity. ANSI/IEEE Std. 176–1987, 1988.

[84] N.S. Ottosen and H. Petersson. *Introduction to the Finite Element Method*. Prentice Hall, New York, 1992.

[85] H.J. Pain. *The Physics of Vibrations and Waves*. John Wiley, Chichester, 1983. 3rd ed.

[86] M. Pawlak and E Rafajłowicz. On restoring band-limited signals. *IEEE Transactions on Information Theory*, 40(5):1490–1503, September 1994.

[87] M. Pawlak and U. Stadtmüller. Recovering band-limited signals under noise. *IEEE transactions on information theory*, 42(5):1425–1438, September 1996.

[88] R. Pintelon, P. Guillaume, Y. Rolain, J. Schoukens, and H. Van Hamme. Parametric identification of transfer functions in the frequency domain. *IEEE Transactions on Automatic Control*, 39:2245–2260, November 1994.

[89] R. Pintelon, J. Schoukens, and H. Chen. On the basic assumptions in the identification of continuous time systems. In *Proc. IFAC Symposium on System Identification, Vol*, pages 143–152, Copenhagen, Denmark, July 1994.

[90] E.P. Popov. *Mechanics of Materials*. Prentice Hall, New Jersey, 1976. 2nd ed., collaboration with S. Nagarajan and Z.A. Lu.

[91] H.R. Pota and T.E. Alberts. Multivariable transfer functions for a slewing piezoelectric laminate beam. *ASME Journal of Dynamic Systems, Measurement, and Control*, 117:352–359, September 1995.

[92] A. D. Poularikas. *Handbook of Formulas and Tables for Signal Processing.* CRC Press, 1999.

[93] C.T.F. Ross. *Finite Element Programs for Structural Vibrations.* Springer-Verlag, London, New York, 1991.

[94] M. Rotunno and R. A. de Callafon. Fixed order H_∞ control design for dual-stage hard disk drives. In *Proc. IEEE CDC*, Sydney, Australia, 2000.

[95] I. J. Schoenberg. *Cardinal Spline Interpolation.* PA: Society of Industrial and Applied Mathematics, 1973.

[96] N. K. Sinha and G. P. Rao. *Identification of Continuous Time Systems.* MA: Klewer, 1994.

[97] S. Skogestad and I. Postlethwaite. *Multivariable Feedback Control: Analysis and Design.* John Wiley & Sons, 1996.

[98] G. C. Smith and R. L. Clark. A crude method of loop-shaping adaptive structures through optimum spatial compensator design. *Journal of Sound and Vibration*, 247:489–508, 2001.

[99] J.W. Smith. *Vibration of Structures: Applications in Civil Engineering Design.* Chapman, London, 1988.

[100] C. C. Sung, V. V. Varadan, X. Q. Bao, and V. K. Varadan. Active control of torsional vibration using piezoceramic sensors and actuators. In *Proc. of the AIAA/ASME/ASCE/AHS/ASC 31st Structures Structural Dynamics and Matrials Conference*, pages 2317–2322, 1990.

[101] R. Szilard. *Theory and Analysis of Plates: Classical and Numerical Methods.* Prentice Hall, Englewood Cliffs, New Jersey, 1974.

[102] J. B. Thomas and B. Liu. Error problems in sampling representation. *IEEE Int. Conv. Rec. (USA)*, 12(5):269–277, 1964.

[103] S. Timoshenko and D.H. Young. *Elements of Strength of Materials.* Van Nostrand, Princeton, 1968. 5th ed.

[104] V. Toochinda, C. V. Hollot, and Y. Chait. On selecting sensor and actuator locations for anc in ducts. In *Proc. IEEE Conference on Decision and Control*, pages 2593–2598, Orlando, Florida, USA, 2001.

[105] V. Toochinda, T. Klawitter, C. V. Hollot, and Y. Chait. A single-input two-output feedback formulation for anc problems. In *Proc. American Control Conference*, pages 923–928, Arlington, VA, 2001.

[106] H. S. Tzou. Integrated distributed sensing and active vibration suppression of flexible manipulators using distributed piezoelectrics. *Journal of Robotic Systems*, 6(6):745–767, 1989.

[107] M. Unser. Splines, a perfect fit for signal and image processing. *IEEE Signal Processing Magazine*, 16(6):22–38, November 1999.

[108] M. Unser, A. Aldroubi, and M. Eden. Polynomial spline signal approximations: Filter design and asymptotic equivalence with shannon's sampling theorem. *IEEE Transactions on Information Theory*, 38(1):95–103, January 1992.

[109] M. Viberg. Subspace-based methods for the identification of linear time invariant systems. *Automatica*, 31(12):1835–1851, 1995.

[110] N. Young. *An Introduction to Hilbert Space.* Cambridge University Press, 1988.

[111] Y.Y. Yu. *Vibration of Elastic Plates: Linear and Nonlinear Dynamical Modeling of Sandwiches, Laminated Composites, and Piezoelectric Layers.* Springer, New York, 1996.

[112] K. Zhou, J.C. Doyle, and K. Glover. *Robust and Optimal Control.* Prentice Hall, Upper Saddle River, NJ, 1996.

Index